The
LOST WORLD
of the
DINOSAURS

The
LOST WORLD
of the
DINOSAURS

UNCOVERING *the* SECRETS
of the PREHISTORIC AGE

ARMIN SCHMITT

WITH ILLUSTRATIONS BY BEN RENNEN

HANOVER
SQUARE
PRESS

HANOVER
SQUARE
PRESS™

ISBN-13: 978-1-335-08121-6

The Lost World of the Dinosaurs

First published as Großartige Giganten in 2023 by dtv Verlagsgesellschaft mbH & Co. KG, München. This edition published in 2024.

Copyright © 2023 by dtv Verlagsgesellschaft mbH & Co. KG, Munich

Illustrations by Ben Rennen

Hanover Square Press
22 Adelaide St. West, 41st Floor
Toronto, Ontario M5H 4E3, Canada
HanoverSqPress.com

Printed in U.S.A.

For Elisabeth and Maximilian

The
LOST WORLD
of the
DINOSAURS

TABLE OF CONTENTS

PROLOGUE

Since the age of five, I have been fascinated by the giant dinosaurs of prehistoric times. I am very grateful that my passion and purpose were revealed to me so early on and that I was able to turn this enthusiasm for dinosaurs into my profession. Today, when I find a dinosaur bone in the field, I still have the same glow in my eyes as I did as a small child when I first saw real skeletons of these giants at the Senckenberg Natural History Museum in Frankfurt. The path from dino-crazy child to vertebrate paleontologist, however, was not straightforward by any stretch of the imagination. Nevertheless, I had the great fortune to be part of many discoveries, some of them very important, even if I myself was not at the center of the research. Some research projects I was involved in contributed to a better understanding of the evolution of dinosaurs and other prehistoric reptiles and fundamentally changed our view of dinosaurs. Some of that is reflected in this book. It's a great feeling to be at the heart of research in the golden age of paleontology, using cutting-edge technology to uncover new mysteries and unravel puzzles that just a few years ago were thought never to be explained. One reason we've recently learned so many new things about dinosaurs is the incredible rate at which new dinosaur species are being scientifically described. In the last twenty years alone, more dinosaur species have been discovered per year than ever before. Since 2003, about forty-five new dinosaurs have been discovered each year. This is mainly due to increased funding for research, more scientists today, and more teams in the field. Remote sites are more accessible, and we have increasingly accurate geological maps with precise coordinates for these sites. It is not uncommon for new fossils to be discovered at already known localities, and it is not uncommon for this to bring new species to light. Also partly responsible is the rapid increase in computing power and speed of computers and workstations, which make it possible to generate simulations and three-dimensional models and to evaluate complex statistics to generate phylogenetic trees of dinosaurs. And recently discovered dinosaur species not only fill museums and provide new names that are difficult to pronounce, but also provide insights into how different animals were related as well as their habitats and survival strategies. Each new find gives us ever-more-precise insights into the ecosystem of the dinosaurs, and their external appearance even helps us better understand their internal structure and the function of their organs. This is enormously important because some dinosaurs had such unusual anatomies that we cannot find anything comparable in nature today. New discoveries help us understand how dinosaurs were able to cope with drastic climate changes and adapt to new habitats. Dinosaurs were fascinating creatures that can serve as model organisms to understand many mechanisms of nature and evolution. In this book, based on several prominent examples of species of dinosaurs, I discuss the fact that there are always new discoveries to be made and that, despite our extensive knowledge of dinosaurs, many questions remain unanswered.

THE GREAT DYING

When the Earth was burning and the sea was choking

To understand how dinosaurs became the most successful terrestrial vertebrate group in Earth's history, we must travel to the beginnings of dinosaur evolution and clarify what factors were necessary for dinosaurs to begin their triumphant reign.

Most children today know that all dinosaurs, with the exception of modern birds, became extinct at the end of the Cretaceous period about 66 million years ago. An asteroid hit Earth at that time and sealed the fate of the dinosaurs and many other animal groups. Ammonites disappeared from the oceans. These were cephalopods related to today's *Nautilus* and squids. They lived in our planet's oceans for well over 300 million years and were an important component of complex food webs in marine ecosystems. Along with the ammonites, large marine reptiles such as the four-flippered plesiosaurs and the varanid-like mosasaurs also disappeared without any descendants.

But as terrible as this cosmic impact must have been, it is dwarfed by an event that had much greater consequences and is widely considered the worst catastrophe for life on this planet. It was a cataclysm of apocalyptic proportions, so devastating that some authors refer to it as "the Great Dying": the Permian–Triassic mass extinction.

Michael Benton, one of the most influential paleontologists of our time, wrote a book about it, giving it the striking title *When Life Nearly Died*. This event, which took place at the end of the Permian period, resulted in the greatest mass extinction event in the history of the Earth. About 90 percent of all species living at that time became extinct. For the first time in Earth's history, many insect species were also affected.

During the Permian, food was still plentiful for all animals. The Earth was a green planet. For millions of years, plant growth had been so abundant throughout the world that the oxygen content in the atmosphere continued to rise and reached its highest level in the entire history of the Earth. Today, we assume that the atmospheric oxygen level was around 30 percent in the Permian.

> **By comparison, the oxygen level in the atmosphere today is only about 21 percent.**

Droughts, fires, air pollution, and acid rain subsequently destroyed the lush forests, causing the atmospheric oxygen content to drop from 30 to about 10 to 15 percent and then remain at low levels for many millions of years. Oxygen levels would never again reach those from the beginning of the Permian. The effects were so far-reaching that it took almost 15 million years for the Earth's forests to fully recover.

"The Great Dying," and the subsequent collapse of the ecosystems, is thought to have been caused by mega-volcanism in Russia. The so-called Siberian traps, or flood basalts, are a testimony to this catastrophe. These are extensive lava flows that stretch from the Barents Sea in the northwest to Kazakhstan in the southwest. From there, they continue to Novosibirsk and Irkutsk in the south. In the east, they extend beyond the Lena River almost to Yakutsk. Some of you may recognize Irkutsk and Yakutsk from the board game *Risk*. The volcanic eruptions occurred in three spurts, over a period of about nine hundred thousand years, and covered an area of 7 million square kilometers with volcanic rock. This is equivalent to almost the entire area of Australia!

The volcanic eruptions responsible for this are considered to be among the largest known volcanic events in Earth's history. The eruptions took place about 252 million years ago at the Permian–Triassic boundary and their consequences are causally linked to the mass extinction at the end of the Permian, or "the Great Dying." The postapocalyptic world after the Permian catastrophe must have come very close to our idea of hell. The mega-volcanism in Siberia released substantial amounts of carbon dioxide, fluorine, hydrogen chloride, and sulfur dioxide, which reacted with rain to form sulfuric acid, permanently damaging or destroying both marine and terrestrial habitats. Carbon dioxide emissions triggered global warming, increasing the global temperature by 9°F in a very short period of time. We now also suspect that the ozone layer was severely damaged by the greenhouse gases and the fly ash, resulting in an increase in UV radiation, in turn hampering plant growth and causing pollen malformations. Massive forest fires accelerated the large-scale vegetation loss, and acid rain finished off most of the plants. At that time, it only rained periodically, during the mega-monsoon season. But then, the celestial floodgates would open wide and pour forth torrents of caustic rain for days and weeks across the supercontinent of Pangaea. The rest of the year was marked by terrible heat waves, droughts, fires, and an unbearable average temperature of 113°F. While there was still sparse vegetation in the coastal regions, the interior landmass was dominated by vast deserts wider than even the Sahara or Gobi deserts. The consequences were devastating. The extreme environmental conditions increased erosion on land. This resulted in extensive draining of the soil into the sea, leading to eutrophication effects such as algal blooms in marine habitats. At the same time, there was a mass proliferation of marine protozoa in oxygen-free environments, whose metabolic products—in addition to all the gases from the massive volcanism—released methane, halocarbons, and large amounts of hydrogen sulfide (H2S) into the atmosphere.

In the meantime, anoxic—oxygen-free—conditions developed in the oceans, causing all reefs and their inhabitants to die off. Rocks that bear witness to this momentous event are called black shales, which are dark-colored organic-rich sediments. With the formation of anoxic marine zones, the rapid decrease in pH, and the release of methane hydrate, a mass extinction began in the oceans. As a result of the methane escaping into the atmosphere, the Earth's temperature rose by a further 9°F and the upper water layers of the world's oceans even warmed by at least 14°F. The Earth was burning and the oceans were boiling. The ever-increasing concentration of greenhouse gases led to a domino effect, which we describe as a galloping greenhouse effect, that explains why the mega-volcanism in Siberia lasted almost a million years but the mass extinction only thirty thousand. Initially, the changes were gradual, but as more and more habitats were damaged and more and more pollutants were emitted, temperatures increased faster and faster, and the collapse of the ecosystems became unstoppable. In the end, about 95 percent of all marine species and about 75 percent of all terrestrial species became extinct during this event. Among the few survivors were the direct ancestors of dinosaurs and crocodiles: the so-called archosaurs.

Without this catastrophe, the triumphant rise of the dinosaurs would probably never have happened. And so it is perhaps an irony of fate that, on the one hand, the dinosaurs entered the world stage after a devastating mass extinction, while on the other, they then fell victim to a mass extinction themselves. There have been five such events of mass extinction in the history of the Earth. Three of them are closely related to the origin of the dinosaurs and were crucial for their unprecedented evolutionary success—but ultimately also responsible for their demise:

1. The Permian catastrophe about 252 million years ago ("the Great Dying") favored the development of the archosaurs. The archosaurs, or "ruling lizards," include all crocodiles and dinosaurs. Birds also belong to this group, but they did not yet exist in the Triassic. The archosaurs, and dinosaurs in particular, were able to hold their own well in the hostile world of the Triassic because their efficient lungs were better able to cope with the low levels of atmospheric oxygen, they could run faster with their long, straight legs, and their bodies were farther from the hot ground than those of other lizards and amphibians. These traits were advantageous in the oftentimes desertlike regions of the prehistoric continent. This allowed them to colonize new habitats. Their scaly skin prevented their bodies from drying out, which was an invaluable advantage in the sweltering heat of the Triassic.

2. The end-Triassic mass extinction, which occurred about 201 million years ago. This catastrophe killed the conodonts in the sea, which had existed for about 340 million years and now disappeared without descendants. As either the first vertebrates, or close relatives of today's vertebrates, they are among the earliest forms that later gave rise to fish, amphibians, reptiles, mammals, and birds. Further impacting dinosaur evolution was the extinction of four major groups of crocodilian relatives at the end of the Triassic: rauisuchians, phytosaurs, aetosaurs, and ornithosuchians. These

representatives of the crocodilian lineage were similarly as successful as the dinosaurs in the Triassic. At the same time, almost all temnospondyls, a clade of giant amphibians that could grow up to thirteen feet long, became extinct. They are considered a sister group of today's frogs and salamanders. For simplicity, they can be thought of as thirteen-foot-long monster salamanders with bony plates on their skin. With the demise of the monster salamanders and most land-dwelling crocodiles, there was no longer any competition for the dinosaurs, and so they became the undisputed rulers of the Earth at the beginning of the Jurassic.

3. The end–Cretaceous mass extinction about 66 million years ago finally sealed the fate of all dinosaurs that were not true birds and not on the bird lineage. The end of this book explores why the dinosaurs did not survive the asteroid impact.

> The fossil record includes all scientifically documented finds of fossils worldwide. The occurrence of fossils at a certain place is usually mentioned in scientific publications. Therein, the geographical location as well as the stratigraphic context of the finds are given. Investigating, adding to, and ultimately completing the fossil record are some of the most important tasks of paleontology.

Between these three catastrophes stretched incredibly long periods of time. The evolution of the dinosaurs took place between the Permian catastrophe and the impact of the asteroid at the end of the Cretaceous period. These events are about 186 million years apart. In order to imagine such a time span, one must keep in mind that *Tyrannosaurus* and humans are separated by a little more than 65 million years. The last *Tyrannosaurus* became extinct 66 million years ago and humans (*Homo sapiens*) emerged about three hundred thousand years ago. *Stegosaurus*, on the other hand, became extinct almost 148 million years ago, so it lived about 80 million years before *Tyrannosaurus*. So, in terms of geologic time, these two are further apart than *Tyrannosaurus* and humans. But the very first dinosaurs emerged another 90 million years or so before *Stegosaurus*. Consequently, *Plateosaurus* (an ancestor of the later long-necked dinosaurs) lived in a completely different world than *Stegosaurus*, and *Tyrannosaurus*, in turn, lived in a completely different world than *Stegosaurus*. The reign of the dinosaurs lasted so incredibly long that entire continents broke up, seas opened and closed again, and entire oceans were created during it.

This thrilling journey of the dinosaurs was sandwiched between two of Earth's greatest mass extinction events. In both cases, these were global disasters. Why is this important? Dinosaurs were the most successful land creatures in the long history of the Earth. No other group of animals reigned on land as long as they did. Yet, at the end of the Cretaceous period, they fell victim to a mass extinction triggered by changes in climate and vegetation. While the cause of this extinction was the asteroid, it only killed a few dinosaurs directly. Countless more animals died in agony in the weeks and months following the impact, after the sky darkened, temperatures changed, forests died, and the food chain collapsed.

This scenario sounds familiar. Today, man is certainly the most successful complex organism on Earth. Just like the dinosaurs, we have colonized every continent, hold our own against every enemy, and use all of the Earth's resources—at the same time, the fastest climate change we know of in all of Earth's history is taking place around us, while we face unprecedented species extinctions, the scope of which cannot yet be foreseen. According to a 2019 estimate, the number of birds in North America has declined by about 3 billion since 1970, or about 30 percent of the population there. In Germany, according to the Nature and Biodiversity Conservation Union, or NABU, the number of breeding pairs of birds declined from 97.5 million in 1998 to 84.8 million in 2009, a decrease of 15 percent in just eleven years. Never before in Earth's history have so many birds died. At the same time, the number of flying insects in Germany's nature reserves has declined by an extremely alarming 76 percent over the past twenty-seven years. But while we might be glad that there are fewer mosquitoes at night, flying insects are the main pollinators of most crops and at the same time an important contributor to the diet of most bird species.

The causes of climate change at the end of the Cretaceous period are different than today, but that does not make the current situation any less severe. The extinction of the dinosaurs had extrinsic causes, while anthropogenic influences are responsible for our current situation. Humans themselves are driving climate change and species extinction.

I often hear that some people don't believe in climate change, and don't believe that humans have any significant impact on climate change. The devastating species loss we are currently experiencing makes surprisingly few headlines. Yet climate change and species extinction can be observed in nature. They can be objectively measured and proven with reproducible scientific experiments, whether you believe in them or not. Personal views and opinions are irrelevant here. And if we choose to ignore the warning signs, the problems will not go away—they will only get worse.

THE TRIASSIC

(251.9 to 201.3 million years
before present)

Cymbospondylus youngorum

Omphalosaurus

Ichthyosaurs

Thalattoarchon

Ammonites

Nautilus

Bivalves

CHAPTER 1

New Life in the Sea

Omphalosaurus: *Crushing It!*

It is morning on Monday, September 26, 2011, and I am climbing up to our dig site from base camp at the foot of the Augusta Mountains in Nevada, as I have been doing for several days. It is about sixteen hundred feet above the camp and there are no roads or footpaths leading to it, only a trail left by wild mustangs. Today the ascent is particularly arduous, because I'm carrying a bag of plaster up the mountain in addition to provisions, photo equipment, and 1.25 gallons of water. The first few miles we could still climb in the shade of the mountain, but by the time we reach the first plateau, the hot sun burns down on us. There is no more shade, because there are no trees growing up here, anywhere. Since it is too dry in the valley, only sagebrush and grasses grow, and their barbs stick to stockings, shoes, and clothing and irritate the skin. In the Augusta Mountains, juniper bushes do not appear until about five thousand feet above sea level, and well above sixty-five hundred feet, almost at the summit of Cain Mountain, grow scattered bony, gnarled pinyon pines and the so-called Mormon tea (ephedra). This phenomenon is called the inverse tree line. In the Alps it is the other

way around—there, the trees grow in the valley and the higher
you go up, the rarer they become, until—between fifty-nine
hundred and seventy-two hundred feet above sea level—they
disappear altogether.

From the first plateau you can still see our camp and commu-
nicate with it via walkie-talkie. In 2011, there was no cell phone
reception here. But you had no electricity to charge your cell
phone anyway, to say nothing of running water. We named this
plateau Red Nose Point because the ground here has a reddish
color. Supposedly, every now and then a mountain lion shows
up in this area, but except for their dried droppings, we didn't
come upon any; sadly, the farmers hunt these beautiful animals,
despite the fact that they don't need even to protect any cattle.

I can feel the strenuousness of the last days in my bones, because I have carried shovels and pickaxes up the mountain, and I
secretly wonder why I do this to myself in my spare time. Sure,
I know our plan—we want to excavate "fish lizards," so-called
ichthyosaurs. But you don't have to travel halfway around the
world to do that. There are ichthyosaurs in England and Germany, too. In Holzmaden, southeast of Stuttgart, lies one of the
most famous sites of these prehistoric animals, right next to the
motorway. So why climb steep mountains in the North American desert? Why all the drudgery? And why are these "fish
lizards" found on a mountain at all? This is one of the most exciting mysteries of vertebrate evolution, as I will explain below.

The rocks of the Augusta Mountains were deposited at the
bottom of a vast ocean nearly 250 million years ago, and enclosed
within them lie the remains of the first fish lizards. And although
paleontology has known about ichthyosaurs for about two hundred years, we still don't know exactly how they evolved. The
first complete skeleton of an ichthyosaur was found in 1811 by
twelve-year-old Mary Anning in Lyme Regis, on the south coast
of England. That was thirteen years before the very first dinosaur was scientifically described. We do know that ichthyosaurs

were reptiles and that their ancestors lived on land—but who those ancestors were is still a mystery today.

What makes solving this mystery so difficult is the highly derived skeleton of ichthyosaurs. The shape of their bones changed greatly during their evolution, and as a result of their extreme adaptation to life in the sea, they differed significantly from their land-dwelling ancestors. They no longer possessed hands or feet, but rather finlike limbs, and their head was long with a pointy snout, achieving the best possible aquadynamics. The neck region was short, and the occiput was immediately followed by the shoulder girdle. These animals were perfectly adapted to life in the open sea, and later forms became fast, persistent swimmers due to their tail fin, which resembled that of sharks, and their torpedo-shaped body. Eventually, they moved only their caudal fin when swimming, while the rest of the body remained immobile. Even the lateral fins were no longer used for locomotion, but only for stabilization in the water and for maneuvering. When propulsion is generated solely by a powerful stroke of the caudal fin, this mode of locomotion is called thunniform. Sharks, dolphins, and tuna (the term "thunniform" is actually derived from tuna) move in the same way. However, their similar body shapes are not an indication of a close relationship, but merely indicate similar ways of life and adaptation to the dense medium in which they move. Living in the open ocean forced these animals to adopt the same body shape during their evolutions because water resistance is much stronger than air resistance on land. The evolutionary pressure toward a streamlined shape is particularly strong in marine animals, which is why animals in the ocean resemble each other more closely. When completely different animals evolve toward the same body type, it is called convergent evolution. The physique of the ichthyosaurs was so strongly influenced by their life in the sea that it was no longer comparable with that of their ancestors, which is why it's difficult to identify and find these ancestors—not least because, since they lived on land, they do not occur in the same sedimentary rocks as their swimming descendants.

The origin of ichthyosaurs is particularly fascinating and mysterious because we find these animals already in rocks that are 250 million years old. There are no finds in the Paleozoic, before the Triassic, but about a million years after the most devastating mass extinction the world has ever seen, the ichthyosaurs suddenly appear—abruptly and en masse! In China and Japan, in Spitsbergen and here in Nevada, we see hundreds, maybe even thousands of fossils of these marine reptiles in the fossil record. Where they came from, no one really knows. But these animals are all completely adapted to life in the sea. Now, a million years may sound like an unimaginably long time from a human perspective, but considering the long evolutionary history of life on our planet and the severity of the consequences of the mass extinction at the end of the Permian, it's no more than the blink of an eye. And in that relatively short period of time, their skeletons completely transformed, their hands became fins, and they evolved from egg-laying land dwellers to viviparous deep-sea swimmers. In the Augusta Mountains of Nevada, we find some of the oldest ichthyosaurs, the likes of which are otherwise found only near the Chinese city of Hefei, about three hundred miles west of Shanghai. The state of preservation of the animals in Nevada is mostly good; bones and skeletons are even preserved three-dimensionally. In Holzmaden, Germany, on the other hand, fossils of the fish lizards were flattened over millions of years as a result of rock compression (diagenesis). The rock beds in Nevada are not completely horizontal, but dip slightly to the south. Thus, as we hike north, from one canyon to the next, we can travel back through time, so to speak, looking at older and older rocks and gradually approaching the Permian–Triassic boundary. The mountain range stretches from north to south, and our excavation takes us into two canyons: the Favret Canyon, whose rocks date from the Middle Triassic, and the Mustang Canyon, farther north, with sediments from the older Lower Triassic.

It's lunchtime, the sun is blazing even hotter from the sky, and I eat my sandwich under a large juniper bush. A small lizard squints enviously from a stone at the sandwich. These little animals are barely larger than sand lizards, but are related to the large iguanas. They have a sand-colored body with brown spots and bright yellow feet. On the neck are two noticeable black bands separated by a white stripe. They're called Great Basin Collared Lizards and I think they're downright pretty. The "Great Basin" in their name refers to the geological structure that includes the Augusta Mountains. It is located in the Western United States and stretches from Oregon in the north to Mexico in the south.

I drink another big sip of water from my gallon, into which I mixed blue Gatorade powder that morning. In the mountain desert here, the air is so dry that you don't even notice how much you sweat, because it evaporates immediately. So, you have to be mindful to drink enough—especially if you do any physical labor here. Back to work. In front of me are eleven large rocks, which are about ten feet across when lined up; together they form a sausage-shaped concretion.

> **A concretion is a mineral aggregate that is often round or spherical. In this case, it is particularly long and narrow, flat and oval. It consists of a hard, fine-grained sediment that was cemented by pore water and grew outward from a crystallization center over time.**

In this concretion, the crystallization center is a marine reptile that we excavated in the days before. It is called *Omphalosaurus* and it is an extremely unusual animal. The concretion measures only about ten feet in length, but since the animal's tail is missing, it could have been about fifty feet long. Despite its enormous size, it was probably not a predator at the top of the food chain but instead fed mostly on hard-shelled animals such as the common ammonites. Ammonites are early relatives of squids, i.e., cephalopods, from whose shells only the head and arms protrude, as in the *Nautilus* still living today. We know the eating habits of

the omphalosaurs because they had a so-called crushing dentition with which they could crush or break open the thick shells
of the ammonites. However, the form and arrangement of the
mushroom-shaped teeth still puzzle paleontologists today—they
are among the most unusual teeth in the entire animal kingdom.

When we finally uncover the animal, we are thrilled. The
fossil is better preserved than it first appeared. In front of us lies
a complete skeleton from this group of animals, better preserved
than anything that has been excavated before. Martin Sander,
one of my professors from the University of Bonn, was at this
very spot a few years earlier and saw a few small bone fragments
sticking out of the rock. Unfortunately, he didn't have enough
time to recover the animal then and couldn't wait to get back
here. He knew immediately what he had found there, because
he is one of the leading experts on this enigmatic group of animals and knows the anatomy of omphalosaurs like no other.

Martin was my teacher and mentor in 2011 and supervised my
thesis. He is a great role model to me and knows the Augusta
Mountains like the back of his hand. The team also included
Lars Schmitz, who was teaching at the University of California, Davis, at the time; Martin's PhD student Koen Stein, who
is now employed at the Free University in Brussels; Herman
Winkelhorst, a Dutch expert on marine reptiles; and me.

Now we are satisfied and lie down in turn, stretched out on
the floor next to the animal, to take photos indicating the scale
of our find before the blocks of rock are plastered. Describing
this moment is not easy for me, because even if not everyone
shares my enthusiasm for ichthyosaurs or dinosaurs, it is certainly
overwhelming for anyone to stand or lie next to a fossil that has
been encased in rock for nearly a quarter of a billion years and
that they have excavated with their own hands. We are the first
people to get a glimpse of this animal. It once sank to the bottom of the sea and was embedded in the mud even before the
dinosaurs roamed the Earth. Since then, entire continents have
drifted thousands of miles apart, oceans have opened and closed,
and massive mountains have piled up and been eroded again.

Plate tectonics have caused the former ocean floor to rise and form into mountains. These tremendous processes straighten out my view of the world and my place in it quite profoundly. A quarter of a billion years—that's incredible!

However, we are not here to break records, but to try to find clues for the solution to one of the greatest mysteries in vertebrate paleontology: Why was there such a complex and diverse vertebrate fauna in the Lower and Middle Triassic seas so soon after the mass extinction at the end of the Permian, and where did these animals come from?

The sediment in which we found *Omphalosaurus* clearly originated from the open sea far from shore. This is also illustrated by the many other ichthyosaur species without crushing dentition that occur in this area. That *Omphalosaurus* was also an ichthyosaur was not clear to everyone. For example, in 2011, Ryosuke Motani, a professor of paleontology at the University of California, Davis, expressed considerable doubt that it belonged to the fish lizards, because *Omphalosaurus* was no ordinary ichthyosaur. Finds of this animal are rare worldwide and almost always fragmentary. *Omphalosaurus* probably did not have the characteristic tail fin of its relatives but still moved sinuously through the water. Because it also had finlike paddles, a streamlined body, and an elongated snout, it was unable to walk on land. However, *Omphalosaurus* fed on hard-shelled animals, not fish like other ichthyosaurs. For a long time, it therefore puzzled researchers trying to determine its origin and affiliation, because finds of this animal mostly consisted only of jaw fragments with bulbous or spherical teeth, whose tooth enamel caps looked like the caps of mushrooms. Characteristic skeletal elements that only occur in fish lizards, and by which *Omphalosaurus* could have been clearly identified as one, were generally absent. Thus, an assignment in the animal kingdom was difficult. These teeth are in fact completely atypical for fish lizards. They could even be easily confused with those of the placodonts from the Germanic Basin. There was a shallow sea at the time of the Lower and Middle Triassic where these lizards were common. They also had a crushing dentition, but their teeth were arranged quite differently. But if only parts of the skull are found and the roof of the mouth is missing, this distinguishing feature may not be present. Examinations with a scanning electron microscope can at most reveal differences in the enamel.

In 2011, the discussion about which animal *Omphalosaurus* exactly was and in which ecosystem it occurred was in any case not yet completely finished. Martin pleaded for an affiliation to the ichthyosaurs and a way of life in the open sea. That day in the Augusta Mountains, the evidence for this was right in front of us. Motani had doubted the affinity to the ichthyosaurs and also saw gaps in Martin's interpretation of the habitat. This had all been difficult for Martin to prove at the time. We did know *Omphalosaurus* from Nevada and Spitsbergen, which was clearly a fully marine signal, but *Omphalosaurus* was absent from rocks of the same age from the Anhui province in China. There'd been no reports of finds there, although those sediments had a similar depositional history. Two finds from the Germanic Basin caused us additional headache, because this was a completely different habitat. There should not have been any omphalosaurs there at all, because it was a shallow sea and other ichthyosaurs were completely absent. Instead, there were placodonts, which are absent from all other sites bearing omphalosaurs, as they probably competed with them for food resources.

The occurrence of omphalosaurs in the shallow sea of the Germanic Basin made little sense, and in almost two hundred years no other omphalosaurs were ever found there except for these two fragments. Many clues also suggest that *Omphalosaurus*'s diet was only found in the open sea, far from the coast. It fed on hard-shelled organisms, all of which were free-swimming, rather than sedentary mollusks or corals, as the placodonts in the Germanic Basin may have eaten. Worldwide, corals recovered very slowly from the Permian mass extinction, and it was not until about 10 million years after the Permian–Triassic transition that coral reefs had fully regenerated. In the shallow sea, this process was somewhat faster.

In any case, there is no evidence of coral or shell reefs in the sediments of Nevada. The sediment consists of sea floor from the open ocean. There is no sign of bur- rowing organisms, such as worms, that would have burrowed through the substrate at the bottom of the sea. Such geobionts are also called benthos, and burrowing, rutting, or digging at or in the ground is known as bioturbation. In this area, how- ever, there was no bioturbation, no benthos, and no corals. The seabed was dead. It consisted only of sulfurous silt with no ox- ygen. Instead, there were myriads of bivalves, called daonellids and cephalopods, swimming freely in the water column.

> **The water column is a defined vertical section of a body of water, extending from the surface to the bot- tom. It serves as a conceptual framework to study the biological features and processes of lakes or seas, in- cluding the presence of micro- and macroorganisms.**

In the water itself, the situation was not much better, be- cause here, too, there were anoxic and oxygen-deficient zones. The sediments that indicate these zones are the so-called black shales. The daonellids and ammonites were apparently able to cope with this oxygen-deficient environment better than other animal species, so much so that this habitat was largely lacking species diversity. However, the few animal species that could cope with the adverse conditions occurred here in abundance.

The anoxic conditions may also explain why we found ten or more ichthyosaurs during our excavation, but comparatively few fish that receive oxygen directly from the sea through their gills. If it were lacking there, they would not survive. Ichthyosaurs, like whales and dolphins, are not dependent on oxygen levels in the sea and breathe atmospheric oxygen after surfacing. Conditions for the fish lizards were less than ideal, but the low oxygen levels in the ocean would not hinder their development as long as they had enough to eat. So, the local story in Nevada seemed clear, even if the mystery of the absence of *Omphalosaurus* in China and its supposed presence in the Germanic Basin still remained unsolved.

Only a study by a team from Bonn led by the paleontologist Tanja Wintrich—who, six years after our excavation in Nevada, examined the two jaw fragments from the Germanic Basin that had been assigned to *Omphalosaurus*—helped to clarify the facts. She used a scanning electron microscope and computed tomography to visualize the complex microstructure of the enamel and the structure of the teeth in the jaws of these two finds. In the case of one piece from the Lower Middle Triassic found in eastern Poland, the characteristics of the tooth-bearing fragment provided unequivocal evidence that the jaw was indeed that of an omphalosaur. This was indicated by the shape of the teeth, the morphology of the enamel surface, the unique enamel microstructure, and the peculiar tooth replacement pattern that was unique to *Omphalosaurus*. The teeth of the animal are indeed amazing. They seem to erupt from the jaw almost at random. Microcomputed tomography shows that at least two more generations lie dormant beneath the functional teeth. However, they do not lie directly under the erupted teeth, but are offset, with no discernible tooth roots, and they do not grow out of sockets, but are simply stuck randomly in the jaw. Because we do not yet fully understand this tooth structure, Tanja always speaks of the enigma that is *Omphalosaurus*.

The other piece from Rüdersdorf near Berlin turned out to be a fragment of the left upper jaw of a placodont. Here the characteristic microstructures in the enamel were missing; the shape and the arrangement of the teeth in the jaw were also different. Now the story made more sense. The shallow part of the Germanic Basin was indeed devoid of omphalosaurs, while the find in eastern Poland was located at a sea gate connecting the Germanic Basin of the Muschelkalk with the open sea of the Tethys. The animal had perhaps strayed or ventured too far into the basin before dying there.

The story was then rounded off when, in 2020, Martin learned that omphalosaur teeth had very well been found in the Anhui province, and quite a while ago, at that. Why colleagues there had not reported these finds and why there are no scientific articles about them is difficult to understand.

But we knew nothing of all this on that Monday in the early fall of 2011. I made a drawing of the position of the individual boulders in relation to each other so that the preparators could rearrange them correctly again later, back in the laboratory. Then I was allowed to wrap the blocks with bandages dipped in plaster and tie burlap around them. This protected them for transport. A helicopter came all the way from the town of Winnemucca for this, to fly them down into the valley in a huge net. Koen and I flew up to the dig site from base camp with the pilot, an elderly Vietnam veteran named Ted. The climb, which usually took us hours, now happened in three minutes. Ted landed the aircraft on a bare, flat spot about three hundred feet from the omphalosaur. He stopped only until Koen and I had hopped out of the cockpit and nimbly stooped to get far enough away under the running rotor blades. Then we had to load the rocks into the net and attach it with a large hook to a rope hanging down from the belly of the helicopter. While it was approaching again from above, we wore earmuffs, yet the noise was still considerable. We waited by the net until the helicopter was close enough. The rotor blades were stirring up dust, and every move had to be executed quickly and precisely so Ted could fly away promptly. It took only a few moments to hook up the net and then away he flew with the precious cargo. We quickly ran to a cliff and we watched the helicopter now flying below us through the Favret Canyon toward the valley. What an impressive sight! We followed the helicopter on foot and descended through the canyon. Halfway down, we passed another prominent spot, which we dubbed Beer Point. From this vantage point we could see the base camp for the first time, and we could already anticipate the cool beer that was waiting for us in a cooler down there. Today we had really earned it.

The concretion with the omphalosaur skeleton being recovered by helicopter. On the ground: myself (left) and Martin Sander (right).

Nevada's Ichthyosaurs: All Just One Big Family

Later, at the base camp campfire, the whole team ate dinner and drank water and a few bottles of beer. Usually, we bought the cheapest beer brand. Once, however, Lars spotted a local beer at the supermarket with an ichthyosaur on its label, and he bought a six-pack of it. It was called *Ichthyosaur India Pale Ale*, or *Icky* for short, and came from Great Basin Brewing Co. It made us wonder why the brewery had a geological structure in its name and offered a beer with a label that featured the state fossil. In the United States, each state has its own official motto, flower, and animal, and sometimes even an affiliated fossil. Nevada's fossil is the ichthyosaur *Shonisaurus*, a giant whalelike fish lizard from the Upper Triassic.

Since our plane was taking off from Reno anyway, we decided we'd stop by the brewery before we left. But for now we were still sitting around the slowly dying fire talking about the day's work and the second big excavation we had planned for the coming days: we were going to dig up the sea monster of the Augusta Mountains that we had discovered on the north flank of the Favret Canyon. This monster was so huge and had such a big mouth that it must have eaten not fish but other ichthyosaurs. At some point, Herman cleared his throat because he felt an itch there. And when Koen and a little later also I felt the same itch in our throats, we knew—it was high time for our "medicine." Lars took out a bottle of Tennessee whiskey, and after each one of us had drunk a good slug of whiskey, the itch immediately subsided. Solely as a prophylactic measure, Martin and Lars also took a sip. It had become dark around us in the meantime, and the clear sky offered a fantastic view of the Milky Way. This place was so secluded and free of light pollution that the stars shone brighter in the sky than anywhere else I had known before. Those evenings around the fire, physically exhausted but satisfied with my day's work, that untouched nature, the camaraderie, and the significant finds make the Augusta Mountains a very special place for me to this day.

The big monster we wanted to excavate was located in the Favret Canyon near where we had previously recovered *Omphalosaurus*, about one hundred and fifty feet farther down the steep slope. The rocks there are from the Anisian, the lowest Middle Triassic stage, and are about 246 million years old. The round dorsal vertebrae of the animal already weathered out of the mountain, just rolling down the slope. The dorsal vertebrae of ichthyosaurs are very distinctive: when viewed vertically from above, they appear circular, and in cross-section they look like an hourglass. We suspected that this huge animal was a *Thalattoarchon*. This was a massive predator, about thirty feet long and with a huge skull. Its pointed teeth were flattened on the sides and measured at least five inches long. In front and back they had serrated edges, with which the animal could cut through the flesh of its prey like a steak knife when biting down. *Thalattoarchon's* full name is *Thalattoarchon saurophagis*, which means "the lizard-eating ruler of the seas." An excellent name, in my opinion.

Thalattoarchon was the very first apex predator of the seas known to science. It could hunt extraordinarily large prey and kill them with its teeth. Its dentition was designed so that it didn't just swallow its prey, as most fish do, but could bite off large chunks of flesh from it. *Thalattoarchon's* teeth were shaped in such a way that they were unsuitable for catching and eating smaller prey. It thus depended on eating other species of fish lizards, as its figurative name suggests. One can perhaps compare its way of life with that of modern-day orcas. Such hunters, which can kill very large prey and are at the top of the food chain, are also called apex predators, and their way of life has far-reaching effects on the ecosystem. For a healthy population of predators to be maintained, prey must be at least a hundred times more abundant than their hunters. Statistically, there may only be about one lion for every hundred zebras in the savannah. In the oceans, the ratio is different from that for land animals, but animals that serve as food for predators must still be available in sufficient numbers before the predator's ecological niche can even be occupied. These niches are referred to as trophic levels.

A trophic level describes the position of an organism within the food chain. The food chain follows a certain order, at the beginning of which are the so-called primary producers, which are organisms that perform photosynthesis and do not feed on other animals. These photosynthetic organisms include land plants and algae or lichens. They are on the first trophic level. Herbivores, such as sheep or cows, are called first-order consumers and are on the next trophic level. They are followed by second-order consumers, which are carnivores that feed on herbivores. Many ecosystems do not go beyond this level of complexity. The second-order consumers are then followed only by decomposers (detritivores) such as molds or bacteria, which decompose dead organic matter and excrement and return their nutrients to the soil and water, making them available to plants.

Thalattoarchon represented at least a third-order consumer—that is, a carnivore that fed on other carnivores. Because only significantly less than 10 percent of one trophic level's energy can be transferred to the next, there are no ecosystems that go beyond five trophic levels. At some point it becomes, energetically speaking, uneconomical to hunt and eat predators because there are so few of them and the chances of a hunting success therefore become smaller and smaller—not to mention that predators can usually protect and defend themselves better than herbivores. Therefore, it is absolutely amazing to find a *Thalattoarchon* in a world that had already developed a complex food web shortly after the mass extinction at the end of the Permian. That would be something like having a super lion that eats other lions. Such things do not occur in today's animal kingdom.

When we stood on this northern slope in 2011, shoveling away many cubic feet of dirt and gravel with a pickaxe and a spade, we did not yet realize how ichthyosaurs could diversify so quickly and occupy so many ecological niches. At that moment, we only hoped to find a skull of the animal as well preserved as possible.

The only specimen of a *Thalattoarchon* recovered up to that point was missing the tip of the snout, but we hoped for better luck. We dug into the mountain and first found the anterior part of the vertebral column, the shoulder girdle, the humeri of the finlike forelimbs, and the occiput of the animal, and it was soon clear that the skull was completely preserved. However, the head, along with its snout, was still stuck horizontally in the mountain. This meant that we had to remove about six or seven feet of debris and overburden and work our way a good ten feet into the hill. We dug with shovels and picks down to a thick layer of whitish-gray tuff and then on to about eight inches above the concretion in which this animal lay encased. At this level, there was a second thin orange layer of volcanic tuff. This served as a useful marker for us in the field. When we saw this tuff layer elsewhere, we knew that the rocks were of the same age and that fossil-bearing sediments lay beneath it, increasing the likelihood of finding more ichthyosaurs. Volcanic tuffs were also important for determining the absolute age of rocks.

> **Tuffs are igneous rocks that are ejected from the mountain during volcanic eruptions and can spread over a large area. It usually consists of volcanic ash and small rock fragments that have been baked together by the extreme heat. We often find zircons that can be used for radiometric dating enclosed within it. If one examines the radioactive isotopes contained within the zircons, which are subject to a**

natural decay process, one can determine the age of these layers very accurately.

Since the fossil we were excavating was directly under a layer of tuff, we were later able to determine its age with a high degree of accuracy in the laboratory. This was important in that other index fossils such as conodonts, bivalves, crinoids, coralline algae, and ostracods were completely absent here—except for the ammonites, whose biostratigraphy can also be used to determine the age of the Triassic. Besides the ichthyosaurs, the ammonites formed one of the groups that recovered most quickly after the Permian–Triassic catastrophe. About one million years after "the Great Dying" they were again abundant worldwide. It is assumed that probably only two genera of these cephalopods survived the mass extinction at the transition to the Triassic, but already in the lowest Triassic more than a hundred genera can be found again. Thus, a very fine biostratigraphic classification of the Lower Triassic is possible.

Biostratigraphy is based on the principle of fossil succession and divides rock units with the help of fossils. According to fossil succession, one fossil community is replaced by another over time. Once a fossil becomes extinct and disappears from the fossil record, it never returns in the stratigraphy.

To our knowledge, however, there was no other record of volcanic ash in the region at that time apart from our tuff finds. So, with these samples, we were able to contribute radiometric data to the abundant biostratigraphic data set for the first time. Therefore, we packed a few large bags full of them and took them with us.

Shortly before reaching the concretion, we had to begin working more carefully. After the thin orange layer, only brooms, brushes, screwdrivers, and oyster knives were used. When we finally uncovered the skull, a huge head about six feet long lay before us, completely preserved up to the tip of the snout, just like the humerus, which measured 16.7 inches. That may not sound very big—but for an ichthyosaur, it's huge.

Now came the most difficult part of our work: with heavy hearts, we had to fill up the hole again. We placed shovels and picks next to the skull, spread a tarp over it, and carefully poured loose gravel on top of it to protect the skeleton from weathering. We knew we would not be able to recover the animal during this field season. The hourly rates for the helicopter were simply too expensive and we did not have enough plaster or a suitable crate to send the monster to Germany. We also didn't have enough time to get the necessary paperwork and logistics sorted. Martin knew, however, that he would be back. We took one last longing look at this magnificent piece, while the suspicion came over us that it might well be an apex predator and a consumer of a third or higher order—but not a *Thalattoarchon*; rather, a completely new species, instead. Indeed, this colossus was also much larger than *Thalattoarchon*. We noted the GPS coordinates and started another fossil search.

Three years later, the animal was recovered by another team, and it took years to carve the fossil out of the rock. It wasn't until 2021, ten years after we first uncovered the animal's skull, that the results were published scientifically. It must have been about fifty-five feet long and weighed about forty tons when it was alive. That's almost twice as long as *Thalattoarchon* and about the length and weight of a sperm whale. The skull alone measured almost seven feet and the jaws were armed with many pointed teeth. This monster was the first giant of the animal kingdom.

It now bears the scientific name *Cymbospondylus youngorum*. The genus name means "boat vertebrae," and the species name honors the couple Tom and Bonda Young, who financially supported the 2014 dig. They are the owners of Great Basin Brewing Co., which also brews the Icky beer. This find is special because this animal is so large yet incredibly old at the same time. While there are similarly sized animals later in the evolution of ichthyosaurs, and some that are even larger, *Cymbospondylus youngorum* is the first whale-sized marine giant in Earth's history. This is so remarkable because it emerged only about 2.5 million years after the very first ichthyosaurs appeared, and its first marine ancestors were only about three feet in size or less. This suggests an incredibly rapid evolution toward gigantism. A similar size increase has taken nearly 25 million years for toothed whales (odontocetes), which include orcas and the sperm whales, ten times as long.

But the rapid rate of evolution is not the only thing that made this animal so special. For a predator of this size to find enough food, there had to be very stable conditions in the food web and prey had to be available in sufficient quantities. In addition, unlike baleen whales, a tooth-bearing animal of this size could no longer feed on small ammonites or fish because it had to meet a much greater energy demand and was no longer nimble and agile enough to pursue small prey. *Cymbospondylus youngorum* therefore probably no longer fed on ammonites or fish—or at most on large ammonites and very large fish—but on other, smaller ichthyosaurs. It occupied a higher level in the trophic pyramid and was probably the top predator in its environment. *Cymbospondylus youngorum* was a third-order consumer—that is, a carnivore that ate other carnivores.

But all this has only recently become known to paleontologists. In the fall of 2011, we continued to search for ichthyosaurs, and Koen found a skeleton that was given the field designation *Koen 1*. This is because during excavations, finds are not initially given a scientific name because it is impossible to be sure what genus or species they are before a detailed examination, and because some finds represent new species of animals or plants that do not yet have a scientific name at all. Martin found another small *Cymbospondylus*, which was named *Martin 1*, and a second one shortly after—*Martin 2*.

Lars came across a small ichthyosaur, a *Phalarodon* (a close relative of the *Mixosaurus*), which was named *Lars 1*.

PHOTO BY LARS SCHMITZ

Lars discovered a slab with ichthyosaur ribs sticking out.

The following day, Martin even found a third ichthyosaur, in whose thorax we discovered another, smaller vertebrae close to its spine, and then two more. The large vertebrae were about two inches in diameter, the small ones only about half an inch. They showed no signs of bite marks, and the bone surface had not been affected by stomach acid. So the big animal had not eaten the small ones, no—we were dealing with a pregnant female. This caused a small sensation! The animal, which we did not call *Martin 3*, but *Martina*, was not only a new species, but also the world's oldest record of a viviparous fish lizard or a viviparous reptile at that time. Prior to this female, only two valid species of *Cymbospondylus* from the Augusta Mountains were known: *Cymbospondylus nichollsi*, which had a body length of about twenty-three feet, and *Cymbospondylus petrinus*, which grew to thirty feet in length. *Martina* must have been about fourteen feet long; later studies of her skeleton showed that she'd been fully grown and sexually mature. Although she was smaller than the already known cymbospondylids, her teeth were just as long, clearly identifying her as a new species.

We spent a lot of time searching the mountainside for bone fragments and vertebrae, which we carefully collected. We used superglue to fill in the cracks and stabilize the remaining articulated pieces. We noted the coordinates of each site and collected any loose pieces. Martin and I worked alone that day because Koen, Lars, and Herman had gone into town to take care of the shipping of the omphalosaur. We covered this site also so that the rest of the skeleton, which was still trapped in the bedrock, was protected and could be excavated during a future excavation. That, too, would be three years in coming. In the end, I shouldered the loose fragments from *Martina*, and we headed back down to the valley.

One reason Martin had taken me on this dig was my hiking experience and what he considered my great resilience. I usually carried the heavy equipment up the mountain and our fossil finds back down. I may not be the most athletic person, but I'm used to hiking, and I'm tenacious and downright stubborn. Even though the equipment was very heavy, I refused to give up. A few months earlier, I had been on an excursion with Martin and other students to the Bletterbach gorge in South Tyrol. There we had found the fossil footprint casts of a para-reptile called *Pachypes dolomiticus* in the riverbed. Such casts are natural infillings, formed when footprints are permeated with harder material that is resistant to weathering. The name of the animal means "heavy foot from the Dolomites." It lived immediately before the Permian–Triassic mass extinction event and was one of the largest herbivores at that time. The casts were quite heavy. But since I already knew that Martin would be traveling to Nevada a little later, I was eager to prove to him that I could handle the arduous marches in Nevada even with heavy baggage. So, I carried the fossil footprint all the way up to the museum, where we handed it over to the curator. I was eager to go with him to Nevada and wanted to show Martin that I was indispensable. None of the other students made such a formidable carrier as I did. Martin actually took me with him to Nevada. We sometimes joked that I was a kind of personal Sherpa for him, but I really wanted to be on the expedition to the Augusta Mountains and was happy to take on all the hardships. So, I carried plaster bags up the mountain and parts of the pregnant female ichthyosaur back down. In return, I was an eyewitness to the discovery of these spectacular finds, saw tarantulas and rattlesnakes, and experienced alpenglow in the Nevada mountains. Once, after a change in the weather where temperatures dropped rapidly, it even snowed on Cain Mountain, and Lars, Martin, and I climbed the snow-covered rise. What a magnificent spectacle of nature!

On one of the last days of our fieldwork, we went even further back into the Earth's past in search of ichthyosaurs in the Mustang Canyon, a little farther north. Martin had never taken on this terrain before because there was just so much to do in the Favret Canyon and never enough time to do it. The Mustang Canyon was composed of Lower Triassic sediments that were even older, going back almost to the Permian–Triassic boundary. If we found ichthyosaurs here, they would surely be the oldest representatives of the group. And indeed—right at the entrance of the canyon—we discovered a place where isolated vertebrae of an ichthyosaur were weathering out of the ground. To our great chagrin, it looked like a fossil hunter had looted this spot and removed the skull. The Augusta Mountains are under the jurisdiction of the Bureau of Land Management, or BLM, an agency within the Department of the Interior, and you can't dig here or take anything without a permit. We possessed such a permit because our excavation was for scientific purposes. All of our fossils, after being prepared and fully examined, were brought back to the United States and given to a museum collection. Fossil poachers, on the other hand, looted sites and then sold the skeletons privately to the highest bidder. How severe the damage to the scientific community is, when fossils are taken out of their geological context and cannot be professionally studied, can only be adumbrated. Whoever had stolen this fossil had prevented the discovery and description of an unknown species.

I think it was Koen who first found the remaining pieces. His sleuthing was remarkable, but we didn't assign a dig number or nickname this time because the sparse remains weren't distinctive or diagnostic enough to make sense of. Our irritation was understandably great. But the carelessly left fragments were followed by *Koen 2*—a small, articulated ichthyosaur—and *Koen 3*, also an ichthyosaur, which actually contained jaws and teeth. These were the first ichthyosaurs from the Lower Triassic discovered in Nevada, and they appeared to be the oldest ichthyosaurs worldwide!

In 2014, however, two scientific articles were published by the aforementioned Japanese colleague, Ryosuke Motani, that changed the status quo of the Nevada findings. In one of these articles, he provided the oldest evidence of viviparity, and in the other paper he reported an ichthyosaur find from the city of Chaohu in China that was even older than the Mustang Canyon specimens. This toppled both *Martina* and *Koen 2* from their thrones.

After having discovered nothing except seashells and am-

monites during the whole expedition, I finally found my first
ichthyosaur in Mustang Canyon: *Armin 1*. It was another small,
articulated specimen that had been deposited in the same strata
as *Koen 2* and *Koen 3*. We documented the finds in our logbook,
protected them with a cap of plaster and noted the coordinates.
Then we carefully covered them with a few shovels of soil. There
was no time left for their recovery because we had to leave the
camp the next day. To this day, these three specimens have not
been comprehensively scientifically examined. Although we
always have to expect that poachers could steal the pieces, the
Augusta Mountains are so remote and the area is so vast that it
is difficult to find the excavation sites without GPS data.

Our campaign was immensely successful. This was partly due
to the experience and sleuthing of the participants, but also because the fossils there are particularly abundant. This is related
to the environmental conditions of the Triassic, where dead
animals that sank to the seafloor were not eaten by scavengers
because of the anoxic conditions there, but instead quickly covered by sediment.

Much of what we discovered during this and subsequent excavations could not be accurately interpreted at the time, and our
records were eventually surpassed. Science just takes its time,
and records themselves are ultimately unimportant because paleontology is about understanding nature and evolution.

Searching for and recovering fossils in the field is tedious and
costly, but the biggest cost in the end is the preparation in the
lab, which also takes the most time. Our dig took place many
years ago, but only now are publications about these finds coming out. An article on the pregnant female ichthyosaur came
out in 2020 and one on the Augusta Mountains monster in
late 2021. The fact that these publications on our Nevada finds
have been so long in coming is largely related to the preparation
of these animals. It is extremely difficult and incredibly time-
consuming because the ichthyosaurs were stuck in concretions.

As mentioned earlier, these aggregates are formed by circulating pore waters sintering out from a crystallized center within an otherwise fine-grained sediment. This makes the nodule very heavy, solid, and particularly hard. The fossilized bones and the surrounding rock matrix of the concretion have almost the same density and cannot be easily separated along the divisional surface between bone and matrix like with many other fossils.

The *Omphalosaurus* was particularly difficult to prepare. Several preparators in Germany and the United States have worked on it in vain. But although the preparation is not yet finished, we can already say a lot about the animal. Until our find in Nevada, no complete skeleton or even nearly complete partial skeleton of an *Omphalosaurus* had been found and scientifically described anywhere in the world. Our find, however, contained almost all of the dorsal vertebrae, so today there is no doubt that *Omphalosaurus* clearly belonged to the ichthyosaurian clade.

The pregnant female was just as difficult to prepare, and it is only thanks to the excellent work of the preparator Olaf Dülfer at the University of Bonn's Institute of Geosciences that we can now admire this animal, freed from its stone prison. It takes skill and a great deal of experience to prepare such a fossil so that it can be examined in detail. Because of this special achievement, the animal was finally named *Cymbospondylus duelferi* in honor of the preparator.

The exact phylogenetic analysis of all ichthyosaurs from the Augusta Mountains ultimately showed that they had all been closely related and were basically only larger or smaller *Cymbospondylus*. So, all the pieces of the puzzle finally fit together here, too: *Cymbospondylus* was apparently one of those ichthyosaurs that successfully occupied the predator's niche in the Middle Triassic seas because they had no competition from other animal groups. The descendants of this generalist rapidly diversified and colonized different habitats, where large, medium, and small forms fed on different prey and rapidly split into new species. Because no other animals competed with it, *Cymbospondylus* was able to produce specialists in a wide variety of niches.

So, the pregnant fish lizard lady was a new species of the cymbospondylian genus (*Cymbospondylus duelferi*), and the giant monster turned out to be a *Cymbospondylus* of another new species (*Cymbospondylus youngorum*). Their close relationship explains why the ichthyosaurian fauna in Nevada was able to diversify so much in such a short time, and how a highly complex food web was able to reemerge so quickly after the mass extinction.

The smaller ichthyosaurs from Mustang Canyon have not yet been fully prepared, but it will be exciting to find out if they are related to the cymbospondylids. This would bring us a little closer to the origins of this species-rich family. It is almost certain that they are previously unknown species. Whether they all belonged to the same or different species will soon become clear. With their exploration, we are moving one step closer to the mysterious origin of the ichthyosaurs.

Of Strange Monsters and Underwater Flight

In Europe, in the Germanic Basin, the underwater world of the Middle Triassic looked quite different. Here reefs recovered from the mass extinction somewhat faster, and the continental shelves were soon full of life again. The strangest organisms bustled about, both on the beaches and in the water. Today, remains of these unusual creatures can be found on the German-Dutch border in a quarry near Winterswijk. There are sediments containing land creatures that came to the beach at low tide and went in search of food. Their visits were so numerous that the slabs with their tracks are often called "the dance floor." We find traces of an extinct group of herbivorous reptiles called rhynchosaurs, and of chirotheria, the ancestors of crocodiles, whose five-toed tracks look like human handprints, which is where their name comes from (*chirotherium*, "the hand animal"). In this quarry, there are also footprints associated with another group of crocodiles, the aetosaurs. My favorite track from Winterswijk is called *Procolo-phonichnium*. It belongs to an extinct group of reptiles that have scales on the soles of their feet. The limestone in which they are preserved is so fine-grained that even such delicate structures have been preserved. If you take a flashlight in dim light and hold it close to the surface at an acute angle, you can produce grazing light and see the delicate scales particularly well. Because the deposits in Winterswijk are of both terrestrial and marine origin, we find countless footprints from land animals, though no bones; it's a different story for marine creatures. In the sea, we see much more unusual animals, which appear in the fossil record only for a short time during the Triassic and then disappear again. There are, for example, small marine archosaurs called helveticosaurs, or the placodonts ("tablet teeth"), which, like *Omphalosaurus*, had a crushing dentition, but stayed in coastal regions and fed on sessile hard-shelled animals. Other groups included the nothosaurs and—one of the strangest animals of all—*Tanystropheus*. It had a greatly elongated neck, which was as long as its torso and tail combined but consisted of only twelve or thirteen vertebrae.

All these animals are also found in sediments of the same age on Monte San Giorgio. This mountain is located on the border between Switzerland and Italy, and there is a phenomenal view over Lake Lugano from its summit. On Monte San Giorgio, you can find, among others, the pachypleurosaurs, or "thick-ribbed lizards." They belong to the group of "lizard flippers" (Sauropterygia), which were still small in the Middle Triassic, but soon grew larger and larger, rivaling the ichthyosaurs. The most successful group within the lizard flippers was the plesiosaurs. They were extremely diverse and one of the longest-lived marine reptile groups. Plesiosaurs had a unique and enigmatic body shape that we do not see in the animal kingdom today. They possessed four pointed, evenly shaped fins, a long, stiffened neck, and a small head. For nearly two centuries, plesiosaurs were thought to have appeared only after the end-Triassic mass extinction in the Jurassic, until skeletal remains belonging to a Triassic plesiosaur were found in Bonenburg near Paderborn,

Germany. The animal was named *Rhaeticosaurus*, "the lizard from the Rhaetian"—the Rhaetian being the uppermost stage of the Upper Triassic, from about 208.5 to 201.3 million years ago. Histological studies indicate rapid growth and increased metabolic rates in plesiosaurs, which are interpreted as an adaptation to swimming and hunting in the open ocean. The existence of *Rhaeticosaurus* supports the hypothesis that an open-sea lifestyle facilitated plesiosaur survival beyond the Triassic.

The extraordinary anatomy of plesiosaurs, combined with their active metabolism, enabled them to use a mode of locomotion unique in the animal kingdom: they were able to use their four uniform fins like wings for what is known as underwater flight. No other group of animals is capable of this. Today, penguins also move forward underwater in this way, but they only use their two wings; their legs and feet are not involved.

PHOTO BY THE AUTHOR

Mounted plesiosaurid Cryptoclidus *in Bonn, Germany, with its flippers in underwater flight posture.*

Whether plesiosaurs really used their fins for underwater flight or for rowing was not clear for a long time. To answer this question, it was not enough to reconstruct only the arm and leg muscles; the pelvic muscles and the muscles of the phalanges that form the winglike fins also had to be considered.

Hydrodynamics studies show that muscles serving fin rotation in particular are necessary for efficient underwater flight. To reconstruct muscle groups of an extinct species, we must first look closely at the muscle attachment grooves on their bones and compare them with those of closely related groups still living today, such as the turtles, crocodiles, and squamates (scaled reptiles such as lizards and snakes). We also need to understand the corresponding muscle functions in sea turtles, penguins, sea lions, and whales to draw reliable conclusions about how the relevant muscle groups work in secondarily aquatic animals. Studies show that plesiosaurs were very capable of wing flapping with the front and rear extremities, while other muscles were able to rotate the front and rear fins on their longitudinal axes during upward and downward strokes.

Plateosaurus

Thrinaxodon
& Broomistega (Amphibian)

CHAPTER 2

New Life on Land

Survival in Hell: The Road to Dinosaur Domination

On land, Earth during the Lower Triassic was characterized by terrible droughts, wildfires, and a global average temperature of an unbearable 113°F. At this time, all landmasses were still united into a single supercontinent before they drifted farther and farther apart over many millions of years. This primeval landmass is called Pangaea. While there was still isolated vegetation on the coasts or in the mountains, enormous deserts larger than the Sahara spread out across the interior of Pangaea.

Despite all of these ecological difficulties, one of the most beautiful fossils I know of comes from this very period of the Earth's history. My colleague Vincent Fernandez and his team found it in Africa in 2013, then studied and scientifically described it at the multinational research facility ESRF—the European Synchrotron Radiation Facility in Grenoble—using microcomputed tomography. The fossil is not only exceptionally beautiful, but also gives us unique insight into the harsh living conditions around 250 million years ago, as well as the strategies animals used to cope with them. It helps decipher the mystery of how some animals survived at all in this harsh, absolutely hostile, downright infernal environment, and why life on Earth was able to persist after this catastrophe. I met Vincent Fernandez at the ESRF while we were doing research together on a project about the evolution of modern bony fish, scanning ribs and jaws from about a hundred species. Vincent's 2013 paper, however, is not about fish, but about said fossil, which shows two animals of different genera. They lie in a burrow, nestled close together, united in death. Exploring the circumstances of how these two different animals died in the same burrow reminds me of a Sherlock Holmes mystery, where meticulous detective work, tiny clues, and clever deductions lead to the solving of a complicated case.

In the harsh environment at the beginning of the Triassic, every day was a struggle for survival. Today, rocks that can tell us about this drama are found in, among other places, the Main Karoo Basin in South Africa, the stratigraphic sequence of which covers a period from before the Permian to the Jurassic. To understand what happened shortly after the Permian mass extinction, we have to look at the rocks from the *Lystrosaurus* Assemblage Zone, a small section of fossil-rich strata within the Karoo rock sequence that were deposited between 251 and 249 million years ago, immediately after the mass extinction event. Vincent's spectacular fossil came from this very zone. In these earliest layers of the South African Triassic, one finds many fossilized bones, but also countless burrows and subterranean structures, which one might imagine as oversized mole tunnels. They are, however, no longer hollow, and are now filled with sediment.

A burrowing lifestyle is a common adaptation in mammals today. It is useful in raising young and serves as protection from predators and extreme weather conditions. In the strata of the

Karoo Basin, we find hundreds of such so-called ichnofossils immediately after the Permian–Triassic boundary.

> **The footprints, skin impressions, tiny boreholes, and burrow structures of extinct animals are called trace fossils or ichnofossils.**

The large number of burial structures in South Africa suggests that a burrowing lifestyle was widespread among numerous tetrapods more than 250 million years ago.

> **Tetrapods are all land-living vertebrates, i.e., amphibians, reptiles, mammals, and birds.**

The period from which these fossils date is characterized by the harsh climatic conditions of the Karoo, and the disproportionate density of ichnofossils reflects the crucial survival strategy of many tetrapods during this time. Sometimes cave fillings still contain the remains of their creators, providing us with important information about the vertebrate ecology of the time. Despite the abundance of burrow casts, however, documented finds of burrowers within them remain rare, obscuring the burrowing lifestyle of many prehistoric animals. From this we can theorize that burrows provided a stable and secure environment for animals of the earliest Triassic. There is evidence that points to seasonal periods of dormancy for their inhabitants, probably caused by resource scarcity from things like prolonged droughts. The variety of cave forms suggests that these were retreats for different animals and were excavated for different purposes. The most common producers of these trace fossils were so-called therapsids. These are amniotes, from which today's mammals evolved.

Amniotes are a large group of vertebrates that includes all terrestrial vertebrates, with the exception of amphibians. Representatives of the amniotes possess the ability to reproduce completely independently from water. Their development occurs in an amniotic sac filled with amniotic fluid, rather than through a fully aquatic tadpole stage. Amniotes include reptiles, birds, and mammals. Humans are amniotes, too.

Therapsids probably originated in the early Permian and, excluding present-day mammals, had their heyday in the Permian and early Triassic, where they were the dominant predators in some ecosystems. Then the first dinosaurs appeared, gradually replacing them at the top of the food chain. The therapsids were menacing predators of considerable size, but they were relegated to a niche existence by the dinosaurs from the Late Triassic onward. After that, for about 150 million years, there were only small to rodent-sized nocturnal creatures that ventured out of the thicket only after sunset, when the big dinosaurs were asleep.

But at the beginning of the Triassic, these animals were larger—and predatory. One of them was *Thrinaxodon*. It belongs to the subgroup of cynodonts, or "dog-teeth," named for their canine teeth, which are similar to those of dogs. They first appear in the fossil record as early as 270 million years ago. At that time, the world was still suited to them. Plants were still growing lushly and providing food for their prey, but as a result of the end-Permian cataclysm, the vegetation was damaged in many places. Wildfires, air pollution, and acid rain destroyed the lush forests, causing atmospheric oxygen levels to drop from about 30 to around 10 to 15 percent. The surviving cynodonts now found themselves in an entirely new world.

Thrinaxodon has now been identified as the main producer of these burrows in South Africa. Sometimes one of these animals lay curled up in its burrow. *Thrinaxodon* was a carnivore with sharp teeth and grew to be about two feet long. Its legs were under its body and it had a rather short tail. Small depressions were discovered on the sides of *Thrinaxodon*'s skull, indicating whiskers, but were not necessarily evidence of full-body fur. The whiskers probably helped the animal orient itself in the darkness of its burrows. Its burrowing lifestyle is interpreted as an adaptation to extremely hot summers. Studies of the bone microstructure of cynodonts indicate that they spent extended periods of time resting in their burrows. Today, we know such resting phases from, for example, bears, which hibernate. But the cynodonts made their burrows for resting during the summer. This aestivating behavior in therapsids is a survival strategy during the extremely hot summer months.

> **Aestivation is a form of summer hibernation in which the metabolism is slowed down, and all activity is suspended or shut down altogether as an adaptation to adverse environmental conditions. It can occur during periods of heat or drought, and under arid climates it can occur seasonally. Estivating behavior still exists today in some moth and snail species.**

After Vincent Fernandez examined the unprepared fossil mentioned earlier using synchrotron imaging, he knew that the rock block contained *Thrinaxodon* (a few parts of the skeleton even peeked out at the surface), but remarkably, the data also showed a second skeleton. It was even more of a surprise when it became clear that this was a completely different species! Vincent and his team had discovered a unique association of two complete skeletons, lying side by side: a cynodont *Thrinaxodon* and an amphibian called *Broomistega*, embedded in the same sandstone grave. Here, one must think of the *Broomistega* as an oversized, potbellied salamander. Why the two animals were in the same burrow together was completely unclear at first. But, for the first time, this material provided the opportunity to study behavioral interactions between two unrelated Triassic animal species. The team wanted to determine if this interaction was random or if it offered advantages to one of the animals involved. Synchrotron data showed that the amphibian, a juvenile, had had seven broken ribs but had healed. The healing process provided evidence that the trauma had occurred several weeks before its death and had not been caused by the *Thrinaxodon*. The fractures had not been fatal, but they had certainly severely affected the animal's locomotion and respiration. Both skeletons are exquisitely preserved, in complete anatomical articulation and in natural posture, suggesting that the animals were buried with soft tissues intact, including skin and muscle, and possibly even alive. *Broomistega* was not capable of digging a hole in the ground, and therefore may not have been the original burrow dweller, but only a guest. Amphibians such as *Broomistega* are well adapted to life in water and therefore do not have bones as sturdy as land dwellers. Their body structure was unsuitable to build burrows by themselves. Although there is no clear anatomical evidence that *Thrinaxodon* was an active burrower, it can be assumed, since several animals of this species have already been found in burrows. Thus, *Thrinaxodon* was probably the primary occupant and probably the builder of the small burrow, and *Broomistega* only a visitor. Because the amphibian and the cynodont lie closely entwined in this burrow, it is obvious that they died at the same time and that *Broomistega* was not washed into the hole afterward by chance. Otherwise, the animals would lie behind each other.

A prey-predator relationship is also unlikely, as both skeletons are almost completely intact and neither animal gnawed or bit the other. The *Broomistega*, in addition to its rib fractures, has a bite mark over its left eye that could have been caused by the canine teeth of a cynodont. But the distance between the two holes in its bone is 16.5 millimeters, while the canines of the *Thrinaxodon* in the burrow, however, are only thirteen millimeters apart. So, this cynodont was not the culprit. The *Broomistega* seems to have escaped from an even larger animal before that. Perhaps it was from this encounter that it had its rib fractures, too. The poor animal obviously had a rough life.

Another consideration was that the *Thrinaxodon* may have dragged the large salamander into its burrow to eat it later. Storing food in burrows does occur in the animal kingdom, but it usually does not involve hoarding perishable food. Keeping live animals as food in burrows is rather unusual, especially in hot climates that favor rapid decomposition of the carcass. Given the absence of injuries attributable to this *Thrinaxodon* and the low probability of it hoarding perishable food, the above hypothesis can be safely ruled out. Vertebrates of different genera occur together in burrows in the present animal kingdom only when external factors such as foraging, predator avoidance, or harsh climatic conditions play a role. For example, when one of the two species leaves the burrow, an intruder may chase away the host, or when the host tolerates the presence of the new guest. Especially in harsh climates, there are still cases among terrestrial animals today where guests are tolerated by a host. Such communal living offers advantages in warding off predators and also increases the likelihood of survival of the animals involved through shared vigilance. Nevertheless, such cases remain the exception. But juvenile amphibians alive today have been known to temporarily use burrows as protective retreats. This could also be a plausible explanation for the presence of the *Broomistega* juvenile. The fact that the *Broomistega* was not attacked or injured by the presumed *Thrinaxodon* host also suggests that the cynodont tolerated the intruder's presence or was unable to chase it away. Perhaps the *Thrinaxodon* was already dead or paralyzed. However, if it had been dead before its burrow was flooded, rigor mortis probably would have set in long ago. But the lateral curvature of its spine indicates a mortal agony shortly before drowning in the mud. Thus, *Thrinaxodon* either died shortly before the flooding, and rigor mortis had already released, or its death was related to the flooding. Since the assumption that *Thrinaxodon* died before the flood requires exceptional circumstances, it seems more plausible that the cynodont tolerated the salamander in its burrow. The time of *Broomistega*'s death can be reasonably correlated to the flood and argues for voluntary entry into the burrow. Its curved vertebral posture also suggests that it died shortly before the flood event or during the flood.

In the Early Triassic, the climate was dry, and the drought was interrupted only once a year by a violent monsoon season. During the hottest months the animals became immobile, which is why aestivation can also be used as an explanation for the therapsid skeletons in the Karoo Basin burrows. In particular, the mostly dormant or curled-up position of the animal skeletons supports this assumption. Studies of aestivation in contemporary mammals indicate that prolonged inactivity and fasting can lead to growth arrest, particularly in the arm and leg bones. Arrested growth or growth interruptions can be detected in bone cross-sections by various marks and lines.

> **Lines of arrested growth are referred to as LAGs. Such LAGs can be found in ectothermic animals, but normally they do not occur in mammals. However, LAGs have already been observed in the long bones of several therapsids, some of which were discovered in curled-up positions, indicating their estivating behavior.**

Interestingly, however, bone-histological studies of *Thrinaxodon* show no such LAGs. We must therefore assume that a burrowing lifestyle provided a continuous growth rate for the animal. Burrows mitigated extreme heat, and strong climatic fluctuations were balanced by stable ambient temperatures and constant humidity inside them. This helped these animals regulate their body temperature under extreme climatic conditions, and they were also able to cope more easily with the lack of re-sources. The *Thrinaxodon* was able to survive without summer hibernation in a torpor stage.

> Torpor describes a sleep state that occurs in small mammals or birds and is comparable to a kind of lethargy in which metabolic processes are slowed down to a minimum. The body is kept "on the back burner." The animals usually remain in torpor in a state of body rigidity. While in this state, they can survive for a long time without water or food.

Torpor, which can be self-initiated by mammals under ex-treme weather conditions, consists—in contrast to an uninter-rupted estivation—of several short periods of rest lasting only for a few days each. Lines of arrested growth become more in-distinct or are completely absent during this period, so that they can no longer be detected in the cross-sections of arm and leg bones. Since the growth of the animal does not stop completely, the characteristic LAGs are missing.

The unusual residential community of *Thrinaxodon* and *Broom-istega*, and also the numerous other *Thrinaxodon* specimens in curled-up body postures, suggest that this cynodont genus re-treated into its burrow for periods of dormancy. However, in the absence of histomorphological markers indicating an interruption of bone growth, these periods seem to have been short-lived. Such temporary periods of torpor are a primitive feature in mammals and thus more likely than the specialized metabolism of hiberna-tion. In this mammalian ancestor, a certain plasticity in metabo-lism still existed, meaning that these animals could slow it down when external conditions became unfavorable. Humans are not able to do this. If we do not have enough food or water for a long period of time, we die. The ability to escape threatening climatic conditions via a burrowing lifestyle, and thus to survive a food shortage longer, certainly contributed to the survival of small- and medium-sized cynodont animals in the Permian–Triassic crisis, as did the variability or plasticity of their metabolism.

But when the golden era of the big amphibians and cynodonts was over, small nimble bipedal lizards were already running around the burrow of the odd couple who'd been united in death. These bipedal lizards coped much better with the extreme temperatures. And so, the death of *Thrinaxodon* and *Broomistega* is also symbolic of the demise of their respective animal clades. The ordeal of these unequal co-inhabitants fatefully casts its shadow ahead and also heralds the demise of their closest relatives.

The Dragon Who Wanted to Be a *Tyrannosaurus*

The small bipedal lizards that roamed around the *Thrinaxodon* burrow were the archosaurs. Archosaurs consist of, simply put, all crocodile relatives and dinosaurs.

If archosaurs are considered the last common ancestors of crocodilians and birds and all their descendants, then this group originated in the earliest Lower Triassic. Other definitions do not refer to a pure relationship, but to characteristics in the skeleton that are unique to this group.

> **Such exclusive characteristics within a group are called synapomorphies. Archosaurian synapomorphies include teeth set in sockets deeply rooted in the jaw, a large process on the shaft on the posterior side of the femur, at least two lumbar vertebrae fused with the pelvic bones, a cranial opening in front of the eye and two behind it, and an opening in the lower jaw.**

If one follows this definition, one can find the first representatives already in the Kupferschiefer (German for "Copper

Shale") sedimentary unit from the uppermost Permian, which is about 256 to 259 million years old. Such anatomical features seem unimpressive at first sight, but they made the archosaurs the real superheroes of the Triassic—or possibly, as we'll discuss later, the supervillains.

The cranial opening in front of the orbit (the so-called antorbital window) reduced the weight of the skull, which was relatively large in early archosaurs, similar to modern crocodiles. The mandibular windows probably also reduced the weight of the jaw in some forms and better distributed the bite forces, allowing these animals to bite harder and more efficiently. The large process on the shaft on the posterior side of the femur provided a large surface for muscle attachments. Stronger muscles enabled early archosaurs to walk upright and helped them or their immediate ancestors survive the catastrophic Permian–Triassic extinction event. The upright gait also made it easier for archosaurs to breathe while moving around. It freed them from the so-called Carrier's constraint found in lizards.

Carrier's constraint is a phenomenon where air-breathing vertebrates with two lungs must flex their bodies laterally during locomotion and have difficulty breathing while walking. The sideways flexing expands one lung and compresses the other, forcing the used air from one side of the lung into the other and preventing it from being fully exhaled to make room for fresh air. Lizards, therefore, can only do short sprints and must rest afterwards to breathe deeply.

In the Late Triassic, animals with Carrier's constraint were hunted by bipedal species that did not have this disadvantage. Breathing was especially important in the Triassic because the oxygen content in the air had dropped drastically since the Permian and was only slowly rising again. In this context, too, the archosaurs had a decisive advantage: they possessed birdlike lungs! There are a few important differences in the way lung respiration works in birds compared to other vertebrates:

- Bird lungs maintain a constant volume during respiration.

- They are ventilated unidirectionally, which means that air flows only in one direction at all times.

Contrary to that, in human lungs, breathing air moves back and forth, bidirectionally, as if in a tidal flow. This rhythm is the same in all mammals, with the lungs inflated and compressed with the help of the diaphragm, almost like a bellows. However, to achieve unidirectional flow, various air sacs in the body are inflated and deflated in a complex sequence, like a whole series of interconnected bellows. The lungs, located midway between the air sacs, are supplied with fresh air in one direction during both inhalation and exhalation. The air sacs fill with this fresh air by expanding the chest and the abdominal cavity. The sternum swings forward and downward while the ribs and chest wall move laterally. Exhalation is caused by compression of the air sacs with the help of the skeletal musculature. In a world with little atmospheric oxygen, this was a crucial advantage.

Now unidirectional lungs have also been observed in monitor lizards and iguanas. This could mean that this type of lung evolved before the archosaurs split off from the other lizards. If true, this lung design would have formed as early as the Permian, but there is no fossil evidence for this so far. But it could also mean that this concept evolved several times and independently, i.e., convergently, in the animal kingdom.

Without the bronchial system within the lung tissue, however, iguanas lack a key innovation that enhances the airflow pattern in birds enough to enable them to engage in active flight. This bronchial system was likely favored by natural selection because it facilitated metabolic processes during strenuous activity. And while the unidirectional lung may have evolved as early as the Permian, we don't see this particular bronchial system in the lungs until after the divergence of crocodilians and dinosaurs in the Triassic, which correlates beautifully with oxygen scarcity in the atmosphere. A super-lung, while the other animals were running out of air—this brought the development of the dinosaurs furiously forward and displaced more and more other animals from their habitats.

But not everything was about breathing. To be a successful predator, you also had to have big jaws with strong teeth and be strong enough to kill your prey. So sometimes sheer muscle helped. But if you wanted to hunt large prey, you also had to be able to chomp them into bite-sized pieces after killing them. This meant cutting through bones to separate body parts from the carcass.

The ability to crush or even consume bones as a food source is called osteophagy. Osteophagic feeding behavior, however, seems to have been rare in carnivorous dinosaurs. This can be well explained if we remember that dinosaur skulls were very lightly built. They had one cranial opening in front of the eye and two behind it, and an opening on the lower jaw to optimally distribute and relieve stresses generated during biting. All dinosaurs possessed these openings; they are synapomorphies of archosaurs. In addition, the teeth of predatory dinosaurs were narrow and bladelike. They could cut meat very well with them, but they could not crush bones. These teeth would have usually broken off or would have chipped if the animals had attempted an osteophagic diet. Therefore, we rarely find bite marks on bones in ecosystems dominated by dinosaurs.

The tyrannosaurids of the Late Cretaceous are a major exception. They had massive skulls and thick, robust teeth rooted deeply in their jaws. In these animals, we often see worn teeth in their jaws and deep bite marks on the bones of their prey that match with the shape of the teeth of *Tyrannosaurus*. In the absence of such bite marks, it is difficult for us to say with certainty what the predators ate. Of course, we can use the fossil record to determine which animals might have likely been prey, but that is very imprecise. Here, the fossilized stomach contents of dinosaurs prove much more useful. Unfortunately, fossils that reveal their last meal are relatively rare. In contrast, it is much more common to find the fossilized feces of dinosaurs, the so-called coprolites, through which conclusions can be drawn about their eating habits. The coprolites of carnivores are found much more frequently than those of herbivores because they contain phosphate and therefore have much greater preservation potential. However, in some birds, clues about their diet can be found without having to examine their feces: owls and other birds of prey regurgitate hard-to-digest remains of their prey. These are called spit balls or pellets, which can also fossilize and are then called regurgialites.

One colleague who knows about the coprolites of archosaurs like no other is Martin Qvarnström of Uppsala University in Sweden. I met him at an academic conference in Albuquerque, New Mexico, where he gave a talk about a paper on Upper Triassic coprolites in Poland. He had scanned these coprolites at the ESRF using synchrotron microtomography, and his talk in Albuquerque was impressive. Martin has a very casual yet confident way of talking about his field; he is able to explain complex issues in a simple way and to captivate the audience.

Martin examined bone-rich coprolites and regurgialites from an archosaur whose anatomy was already very similar to that of the dinosaurs. They were found from the latest Middle Triassic and the early Upper Triassic in Poland. This archosaur bears the scientific name *Smok wawelski*. The animal was named after the dragon of Wawel from the Polish folk tale. So, *Smok* has nothing to do with the dragon Smaug who watches over a treasure of gold in a mountain from Tolkien's novel *The Hobbit*. This is not a character from a fantasy novel, but a real predator from the Triassic.

The peculiar skull structure of the animal and the contents

of its coprolites suggest that it was able to crush and consume bone. Some other anatomical features also suggest osteophagy in *Smok*. For example, the Wawel dragon had a massive head and robust body like the tyrannosaurs. *Smok* is thus visually similar to *Tyrannosaurus*, providing evidence of convergent evolution associated with a similar feeding ecology.

The composition of the coprolites of *Smok* confirm what the skull structure already suggests. Half of their content consists of tooth and bone fragments. However, the tooth fragments are not from its prey, but rather broken tips of its own teeth, which it swallowed with its food. It ate a rich diet consisting of fish, small mammals and lizards, giant dicynodontians, and large temnospondyls. The dicynodontians and temnospondyls were so large that *Smok* could not swallow them all at once, and many of the salamander-like creatures had bony skin plates that it had to bite through first.

Another animal from Poland was even more similar to dinosaurs: *Silesaurus*. It belongs to a group that we call the Dinosauromorpha. These are archosaurs that already looked almost like real dinosaurs, but still differed in some features. *Silesaurus* had four long, slender legs and a graceful build. It was a little over seven feet long and is now considered the closest relative of the real dinosaurs. With over twenty of its skeletons discovered, *Silesaurus* is the best-studied dinosauromorph. It has contributed much to our better understanding of the early evolution of dinosauromorphs.

Martin Qvarnström has also studied coprolites from this animal and made exciting new discoveries. *Silesaurus* had long been thought to be an herbivore, but Martin and his team discovered through its coprolites that *Silesaurus* was actually an omnivore. In its fossilized feces, they found the remains of many chitinous carapaces of beetles. The fact that we can make such exciting discoveries in coprolites at all is thanks to computed tomography and powerful synchrotron equipment. In the past, it was necessary to cut open coprolites with a stone saw to see what they contained. From the outside, it is usually impossible to see if and which kind of food remains are hidden inside the coprolites. When cutting the stones open, there is of course a risk of destroying their contents or cutting the coprolite in the wrong place, so that nothing at all can be seen on the cut surface. With microtomography, however, we can see all the details in incredibly fine resolution without having to destroy the fossil.

These scans show us that *Silesaurus* hunted insects. Perhaps it did this by stirring up leaves on the ground and then quickly snapping at them when a beetle appeared. This strategy can still be seen in birds today. Although the coprolites contained beetles, suggesting that *Silesaurus* mostly ate insects, it is likely that it also consumed plants. This is indicated by its teeth, which were distributed very irregularly and less numerous than in the jaws of most other herbivores. But the animal is interesting in other ways, too, because it tells us how the open ball-and-socket joint on the dinosaurs' pelvis evolved, and, in the future, could possibly reveal when archosaurs developed feathers.

The oldest evidence of feathers in the fossil record is about 150 million years old. These feathers belong to the Bavarian "Urvögel," which I will talk about in detail later. It is unclear at the moment if dinosauromorphs already had feathers, because so far, we have no evidence for them. It would be exciting to know at what time the feathers developed, because even if there is no evidence for them so far, it does not necessarily mean that feathers were absent in dinosauromorphs. Studies of the internal bone structure of *Silesaurus* revealed that, just like the dinosaurs, it had a rapid growth rate and a very active metabolism. Thus, it was able to maintain a constant internal body temperature. Feathers would have helped *Silesaurus* better retain this heat.

Silesaurus used to be considered a so-called facultative biped, because it was believed to be able to choose between locomotion on all four or on two legs. Most other dinosauromorphs were believed to be exclusively bipedal. However, the latest research suggests that *Silesaurus* and its relatives were quadrupeds throughout, because their forelimbs were significantly longer than previously thought. This means that bipedalism, or movement on two legs, did not evolve until the appearance of the real dinosaurs. In addition, recent findings show that *Silesaurus*'s limbs looked quite different from those of dinosaurs. Its thigh bones were straight, but they were connected to the sides of its hips. This is different from dinosaurs and mammals. And because the condyle of the femur in *Silesaurus* was neither inclined nor fitted into an open socket, and the legs were not under the body, these animals were not true dinosaurs and consequently were not ornithischian (bird-hipped) dinosaurs, either. In fact, it is the open socket that distinguishes dinosaurs from other animals. This evolutionary step must have occurred only after the emergence of the silesaurs and shows that the first true dinosaurs were more innovative than previously thought. Although *Silesaurus* walked on four legs, its front legs were used more for support and less for locomotion. It was only a small step before the first dinosaurs balanced their bodies so that they could move bipedally. This offered the advantage of being able to run faster than on all fours, and their heads were higher off the ground this way. At the same time, this freed their hands and opened them up for new uses; for example, they could reach for food or use them as weapons. It also explains why the real dinosaurs eventually outcompeted the dinosauromorphs, as their very special, agile mode of locomotion is considered one of the key innovations that helped the group to succeed.

Where the Wild Things Are

The rise of the dinosaurs was an important event in the history of vertebrates. However, the exact time of emergence and the early split of dinosaurs from other dinosauromorphs is not entirely clear. It is often difficult to distinguish between dinosauromorphs and dinosaurs, as we have already seen with the example of *Silesaurus*. But there is an animal from Tanzania where the question of affiliation is even more difficult to answer. It is called *Nyasasaurus* and was first scientifically described in 2013.

The genus name refers to Lake Malawi, which is also called Lake Nyasa. *Nyasasaurus* was either the oldest dinosaur or the most closely related sister taxon of this group.

> In biology, a taxon is a name for a group of living organisms. This group is assigned a specific rank, such as a species, genus, family, or order. The International Code of Zoological Nomenclature governs which scientific name is correct for a particular taxon.

The animal has a unique combination of dinosaur-like characteristics and already shows an increased growth rate. The study of *Nyasasaurus* indicates that the global dispersal of dinosaurs was probably very slow at first. Remains of the animal are from the Middle Triassic and are about 15 million years older than the second oldest dinosaur from that region. A 2013 phylogenetic analysis shows that *Nyasasaurus is* indeed a dinosaur. However, although it is related, it is still outside all other dinosaur groups; it is neither a true saurischian dinosaur nor a true ornithischian dinosaur. But since the skeleton of *Nyasasaurus* is incomplete, further finds could still change the classification, which is why the exact time of the origin of the dinosaurs and the pace of their diversification remain unclear to this day. Nevertheless, there is growing consensus among paleontologists that dinosaurs were already primordial, albeit rare, components of Early Triassic terrestrial ecosystems and that their ascent was slow and, in most cases, regional. The oldest dinosaurs that can be reliably dated and identified are from the early Upper Triassic of Argentina about 230 million years ago. By the end of the Carnian, all three major dinosaur lineages—the Ornithischia, the Sauropodomorpha, and the Theropoda—had evolved. Among the oldest dinosaur finds worldwide are *Eoraptor, Saturnalia,* and *Panphagia* from South America, which have been assigned to the sauropodomorph lineage. The report on the oldest real dinosaur from Africa, *Mbiresaurus*, was only published in 2022. *Mbiresaurus* comes from Zimbabwe.

Sauropodomorpha was the first major dinosaur group to spread and diversify during the Triassic. During this time, many dinosaurs underwent major changes in their body structure, becoming herbivores, increasing in size, and switching to four-legged locomotion. By the end of the Late Triassic, about 30 million years after the dinosaurs emerged, sauropodomorphs dominated the niche of large herbivores in all continental eco-systems worldwide. There were different lineages with widely varying body sizes, from bipedal species weighing about ten kilograms to quadrupeds weighing more than five tons, plus animals with different feeding biomechanics. Currently, there are about fourteen different species of sauropodomorphs from the Upper Triassic strata of northwestern Argentina and southern Brazil. This is about half of all species known worldwide for the Triassic. These finds have significantly improved our knowledge of the sauropodomorph fauna over the past fifteen years. These South American sauropodomorphs are important for understanding the origin and early diversification of this group, showing us what helped the dinosaurs achieve their preeminence within the Late Triassic group of terrestrial vertebrates.

The First Giants:
Hundreds of Long-Necked Dinosaurs Stuck in Mud

The oldest known dinosaurs may have come from the Triassic in South America or Africa, but there are remarkable finds in Europe as well. One of the most interesting and probably largest European sites for dinosaur remains is located in northern Switzerland, near the Rhine River, less than six miles from the German border. Dating from the middle stage of the Upper Triassic, sometime between 228 and 208.5 million years ago, hundreds of plateosaurs died near what would become the city of Frick, and were preserved in the mudstone. *Plateosaurus* was an her-

bivore that could grow up to thirty-three feet long and weigh
up to four metric tons, the largest animal of its time in Europe.

Plateosaurus had a small head that sat on a long, flexible neck
with ten cervical vertebrae. Its body was reminiscent of the large
long-necked dinosaurs (sauropods) of which it was an ancestor. Its
torso was stocky, and it had, similar to its descendants, a long, flexible tail. It carried its long legs vertically under its body. *Plateosaurus*
could walk on all fours, but usually moved bipedally. When doing
so, only its toes touched the ground, but not the soles of its feet,
indicating that it could, at least at times, walk quickly on two legs.
The real long-necked dinosaurs were no longer able to do this.

The biggest difference to the sauropods were the short forelimbs, which were about half as long as the hind legs. With its
short arms and its hands with long claws, it could grasp and hold
large objects. Its chest and shoulder girdle were relatively narrow.

When I visited the outcrop in the active clay quarry in Frick
in 2011 during a field trip, I met a former fellow student from
Mainz who was in the process of recovering a new, nearly complete skeleton of a plateosaur. *Plateosaurus* was probably the most
common dinosaur in the Late Triassic and there are more than
fifty localities in Central Europe where it was found. Since many
complete skeletons of *Plateosaurus* have been found, it is probably one of the best-known Triassic dinosaurs.

The first remains of a *Plateosaurus* were discovered in 1834 by
the chemistry professor Friedrich Engelhardt near Nuremberg.
He passed the find on to Hermann von Meyer, who is now considered the founder of vertebrate paleontology in Germany. Von
Meyer wrote in a letter to the journal *Neues Jahrbuch für Miner-
alogie, Geologie und Paläontologie* on April 4, 1837:

"Dr. Engelhardt in Nuremberg brought to the Assembly of
Naturalists in Stuttgart some bones of a giant animal from a
breccia-like sandstone of the upper Keuper of his region. This
find is of great interest. The bones derive from one of the most

massive lizards, which is related to *Iguanodon* and *Megalosaurus* because of the heaviness and the hollowness of its limb bones. These remains belong to a new genus, which I call *Plateosaurus*; the species is *Plateosaurus engelhardti*. The detail of this I will make known later."

Von Meyer recognized the similarity of *Plateosaurus* to the lizards of the Mesozoic period reported from England. When he named the animal in 1837, the term "dinosaur" did not yet exist—the British paleontologist Sir Richard Owen would only introduce it in 1841. *Plateosaurus* was thus the first dinosaur described in Germany and one of the first ever to be given its own name. At the time, only four other dinosaur genera were known worldwide. When the first finds of *Plateosaurus* were made in the middle of the nineteenth century in Trossingen in southern Germany, Friedrich August von Quenstedt, another contemporary paleontologist, gave *Plateosaurus* the nickname "Swabian lindworm," in reference to the dragons of Germanic mythology.

Although there are more than fifty sites in Germany, and many more in France and Switzerland, most of the complete skeletons and much of the exposed material come from only a few outcrops. The most important three are Frick in northern Switzerland, Trossingen in southwestern Germany, and Halberstadt in central Germany. The depositional conditions of these sites are very similar, which is why the three sites are also described as "*Plateosaurus* bone beds."

These deposits extend over tens of thousands of square feet. In Frick, the bone field is at least one mile long, and a new skeleton appears around every thirty feet. It is interesting that there are a large number of complete *Plateosaurus* skeletons at this site, but that all other vertebrates are missing there. The only exception is a few primitive turtles of the genus *Proganochelys* found here and there among the plateosaurs. The outcrops at Frick, Trossingen, and Halberstadt are particularly intriguing because the skeletons of the plateosaurs there are stuck in the sediment in an upright position. However, the hind legs of many specimens are bent, and the thighs are spread wide so that the knees point to the side. In this position, the legs look a bit like those of frogs. Because of this posture, it was clear from the beginning that the animals, after their deaths, were not further transported on by rivers but buried by the sediment, on the spot. But how did these plateosaurs die? Why are the legs bent in many specimens? And why do we see the same posture in numerous animals at three different locations hundreds of miles apart?

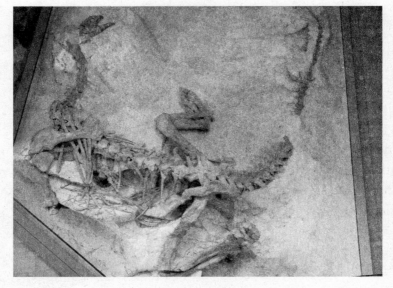

PHOTO BY THE AUTHOR

A Plateosaurus *skeleton squatting in a frog-like posture in Frick, Switzerland.*

The fact that plateosaurs are so abundant in the fossil record indicates that they were the most common herbivores in their ecosystems. They were often found in mass assemblages, indicating that they moved in herds and roamed large areas. The layers where the Frick fossil deposit were found are referred to as playa strata. The word *playa* comes from the Spanish word for "beach." In geology, however, playas are considered plains within mostly drainage-free basins. These basins may be located in valleys between mountains, for example, or they may be coastal plains. Over time, they are gradually covered by layers of saline clay or marl. When it rained or the water table rose, these plains would then turn into shallow salt lakes or salt marshes. For the herds of plateosaurs, especially for the heavy specimens among them that migrated across these playa plains, small, shallow gullies in the dense mud of the marl formed veritable traps. The large animals got stuck in the mire, while the juveniles could escape. The heavy plateosaurs tried to pedal their way out of the mud, but their hind legs could not find solid ground underneath the sticky mud, and, much like in quicksand, they sank deeper and deeper until they finally gave up in exhaustion. The leg muscles were then so weakened that the animals slumped down. The body sank deeper into the mud and the powerless legs were spread apart. While the animals were helplessly stuck in the mud, small predatory dinosaurs came and started chewing on their upper bodies, which were sticking out of the mud, regardless of whether the plateosaurs had already died or were still alive. These predators were not big killers like *Tyrannosaurus*, but actually tiny animals that under different circumstances would never have been able to take down a plateosaur more than twenty-six feet long. Although we had not been able to find any bones of these predators until recently, we knew they were there, as evidenced by hundreds of shed or broken small carnivore teeth in the *Plateosaurus* bone beds. The carnivores preyed on the defenseless plateosaurs in their mud traps, so much so that the upright animals embedded in sediment were often missing their heads, necks, and shoulders.

However, there is a second form of deposition in which complete skeletons in a recumbent position have been preserved. In these specimens, the neck and tail are strongly bent over the back, indicating desiccation and mummification after death. During heavy rains, these mummified carcasses were then carried away by mudflows and redeposited in deeper places. Over time, the mud dried out and settled. This and the load of new sediment flattened the brittle bones of the plateosaurs. The dried soft tissues such as skin, muscles, and tendons were not preserved and they gradually disintegrated.

Although plant-eating plateosaurs and carnivore teeth have been found in the Frick clay pit as early as the 1960s, it wasn't until 2006 that an amateur paleontologist finally discovered the first skeleton of a carnivore in a higher stratum. Unfortunately, this skeleton was missing its head, but that was found three years later during a closer examination of the surrounding rocks. In 2019, the animal was described and given the scientific name *Notatesseraeraptor frickensis*. The genus name *Notatesseraeraptor* means "the predator with mosaic features" and refers to the fact that this carnivore combines different features of different predatory dinosaur clades. The species name *frickensis* refers to the place of discovery, the community of Frick. Although dinosaur remains and their footprints have been found in Switzerland for many decades, *Notatesseraeraptor* is the first carnivorous dinosaur described from this country. Considering that there are hundreds of *Plateosaurus* skeletons from Trossingen, Halberstadt, and Frick, but hardly any other dinosaurs, the discovery of the Swiss *Notatesseraeraptor* seems almost like winning the lottery.

Similar to the predatory dinosaurs, juveniles of *Plateosaurus* are virtually absent from the playa deposits; they were not found until fifty years later at Frick. The smallest specimens measured about sixteen feet. Perhaps smaller animals had been light enough to escape the mud trap of the *Plateosaurus* bone beds. Martin Sander, my former professor from Bonn, even claimed that no young animals would ever be found in Frick. He had led a large-scale excavation campaign there as a young man and contributed significantly to deciphering the causes of the plateosaurs' deaths. For more than twenty years, his prediction seemed to be correct. Then, however, something completely unexpected happened that both disproved and confirmed his prophecy. For her master's thesis, one of Martin's students described several vertebrae of a plateosaur from Frick. The vertebrae were not yet completely fused and could therefore clearly be assigned to a juvenile animal—however, they were huge, much larger than those of some adult animals. Nevertheless, there was no doubt that they were the same species. At first, no one could make sense of it. Then they took a closer look at the internal bone structure of this animal and other plateosaurs. It turned out that these dinosaurs apparently did not yet have the same active metabolism as their descendants. They simply grew quickly when it was warm and there was plenty to eat, and more slowly when there were not enough resources. So, these animals showed a certain developmental plasticity in their body growth. This means that you cannot necessarily tell the age of the animal by its size. This young animal was not yet fully grown but was already quite a bit larger than many of its adult conspecifics. And although it was only a few years old, it was sinking into the mud because its body weight exceeded the critical threshold. So, contrary to Martin's prophecy, there was a juvenile among the victims after all, but this animal was more than sixteen feet long and certainly already weighed more than one metric ton.

It remains to be seen whether Frick will surprise us again in the future with new finds that will force us to rethink our current views. The chances for that are good, because the bone bed seems to extend much farther than previously thought. Only recently, more plateosaurs were discovered during groundwork for a new residential area on the other side of the valley. With the numerous finds of recent years, Frick is now considered one of the most fossiliferous dinosaur sites in Europe. Halberstadt in Saxony-Anhalt will not be able to compete with Frick for this title. Although there were extensive excavations there in 1910 and 1914, the site was later built over and is no longer accessible.

The first dinosaur bones in Trossingen, in Baden-Württemberg, were found as early as 1904. The first excavations began there in 1911, and two complete skeletons found there were then displayed in the Natural History Museum in Stuttgart in 1913. In the 1920s, a second excavation took place, in which German and American colleagues worked together. They recovered a total of twelve skeletons, one of which was exhibited in the American Museum of Natural History in New York. A third excavation took place in the 1930s, but unfortunately, two thirds of the recovered finds were destroyed during World War II.

In addition, there are dozens of smaller sites in Germany, many in Franconia near Nuremberg, where over a thousand individual bones have been found. Other famous sites are in Burgundy, France, and there are even reports of a relative of *Plateosaurus* from Greenland. Even in the North Sea, off the coast of Norway, isolated bones of *Plateosaurus* have been discovered during an oil drilling, eighty-five hundred feet below the seafloor. In the Upper Triassic, *Plateosaurus* was probably widespread throughout Europe, and had closely related genera on every other continent, with the exception of Australia. It is not surprising that these animals were found worldwide, considering that in the Triassic, all continents were still connected and formed the primeval supercontinent of Pangaea. Even in Argentina, relatively large relatives of *Plateosaurus*, such as *Riojasaurus* and *Lessemsaurus*, lived in the Upper Triassic. They were even a bit larger than *Plateosaurus* and, with their length of up to thirty-nine feet, foreshadowed the even larger sauropods that would succeed them in the Jurassic. In the Triassic, however, these prosauropods were still the largest animals in their respective ecosystems. At least that is what was believed for a long time—until an extremely unusual discovery was made recently, not in the field, but in the collection of the Tübingen Museum of Natural History. There, the remains of a "plateosaur" discovered in 1922 were reexamined more closely. It was found that in reality it was not a plateosaur, but in fact a completely new species, much more closely related to the sauropods than to *Plateosaurus*. The new name is *Tuebingosaurus*. This finding could have far-reaching consequences, forcing paleontologists to take another look at the "plateosaurs" in all the historical collections of Germany, France, and Switzerland. Perhaps not all is *Plateosaurus* what appears to be *Plateosaurus*.

At any rate, by the end of the Triassic, dinosaurs had evolved so much that they became the true superheroes of their time.

And so the mass extinction that followed the Triassic offered the dinosaurs infinite opportunities to evolve: a crisis as an opportunity.

The prosauropods evolved into giants and the carnivores became increasingly dangerous predators. The legs they carried under their bodies offered five advantages at once:

1. With two legs swinging vertically under the body, they could run faster.

2. Being bipedal also allowed them to use their hands in other ways. They could grab or hold on to things, and use their claws as weapons to hurt or kill other animals.

3. The upright gait prevented them from dragging their bodies across the ground, as is the case with lizards. This made the dinosaurs less susceptible to overheating, even when the ground was very warm.

4. Their upright posture also enabled them to see farther into the distance and detect potential danger or prey much earlier. This gave them the opportunity to react preemptively, which proved to be a great evolutionary advantage.

5. Their upright posture meant that they did not have to twist their bodies sideways when walking, nor was the air forced

from one half of the lungs into the other. They could exhale it to make room for fresh air. This improved breathing in an environment that was less oxygen-rich than today's.

However, elongated legs extending vertically under the body were not the only key innovation of the dinosaurs. Their avian lungs, combined with more efficient breathing, enabled them to walk with endurance and for long distances. This allowed them to cover vast areas and disperse across the world.

And sometime after splitting off from other dinosaur-like animals, such as *Silesaurus*, the dinosaurs developed plumage. This was perhaps an adaptation to a new catastrophe that struck the world. This time it didn't get hotter, but instead really cold, and the dinosaurs were prepared for it. The end-Triassic mass extinction about 201 million years ago was one of the five major extinction events in Earth's history. In addition to the marine conodonts, it caused the extinction of four major groups within the crocodilian lineage that had previously been competitors of the dinosaurs: the rauisuchians, the phytosaurs, the aetosaurs, and the ornithosuchians. These large groups were similarly successful in the Triassic as the dinosaurs. Yet the causes of mass extinctions during this period are not entirely clear. Some of my colleagues cite sea level fluctuations. In addition, climatic changes must have accompanied the extinctions. Normally, climatic fluctuations are very slow processes, whereas this extinction event may have been quite abrupt. In less than fifty thousand years, all conodonts and many crocodile and amphibian species were extinct. Other possible causes include volcanism or an asteroid impact, or a combination of both.

However, because the sea level was extremely low—it was at a low point worldwide at the Triassic–Jurassic boundary—we must assume that the water was bound elsewhere. In 2022, a study found that there is evidence of moraines at the end of the Triassic in northern China.

> **A moraine is material that is moved along by glaciers then deposited and piled up as debris elsewhere.**

This makes a climate catastrophe seem more likely, and glaciation of the polar ice caps may have led to significant cooling.

The dinosaurs' feathers offered a significant advantage in this harsh environment. This allowed them to access the rich, evergreen Arctic vegetation even in freezing winter conditions, or to hunt for animals that needed these plants for food. Recurrent, violent volcanic eruptions, resulting in volcanic winters with massively reduced solar irradiation, caused mass mortality on land to which all medium-sized to large non-dinosaurs fell victim. These reptiles were not sufficiently insulated and were decimated as a result. They were replaced by dinosaurs that were insulated and protected against the cold. With their plush plumage, these animals had adapted well to cold temperatures and were able to spread rapidly through the Jurassic by adapting quickly. This ecological expansion also led the dinosaurs into regions that had previously been dominated only by large non-feathered reptiles.

THE JURASSIC

(201.3 to 145 million years before present)

Europasaurus

Archaeopteryx

Torvosaurus

CHAPTER 3

The Dinosaurs of the Morrison Formation

The Bone Wars

Everything we know about dinosaurs today is owed to the many paleontologists around the world who are constantly discovering and studying new fossils. What's more, the way dinosaurs are found is often as interesting as the animals themselves. In my view, the most exciting story about dinosaur hunters is about the rivalry between two Americans, Othniel Charles Marsh and Edward Drinker Cope. Among other things, they searched for dinosaurs in the Morrison Formation, one of the most famous rock sequences in North America in which fossils have been discovered, and it was they who first scientifically described *Stegosaurus*, *Brontosaurus*, *Camarasaurus*, and *Diplodocus*. In the nineteenth century, in the so-called Wild West, the two paleontologists engaged in a tense dispute over the discovery of new dinosaur species. There were even shootings and sabotage. The hotheads accepted that some of their finds would be lost to science forever. The feud was fought so bitterly that the country's newspapers reported on it in detail and referred to the dispute as the Bone Wars.

The events of that time were so wild and adventurous that

one might think they had been devised by an author of Wild West novels. Some one hundred and twenty years before the blockbuster *Jurassic Park* drew crowds to the movie theaters, this dinosaur spectacle occupied the media, and people flocked to newsstands in droves to learn of the adventurous developments of the Bone Wars. It was not a Hollywood story that thrilled young and old, but the then-new realization that dinosaurs must have existed. The discoveries of the gigantic bones of the prehistoric lizards appeared on the front pages of newspapers all over the world, and in the United States especially, the interest was enormous. For the first time, reports were published of gigantic, bizarre monsters and their fossils, found in Colorado, Wyoming, and Montana. These areas were almost uncharted in the 1870s, and the army was fighting Indigenous peoples there, pushing them from their ancestral homelands. The stories of dinosaurs, cowboys, Native American tribes, and two eccentric professors were eagerly absorbed by the readers. For not only the finds provided interesting material for stories, but also the people who searched for the primeval lizards. The fact that dinosaurs existed was still new information at the time. The first dinosaur was described by geologist and paleontologist William Buckland in England in 1824, and Sir Richard Owen, the famous British physician and zoologist, coined the term "dinosauria" in 1842, by which time both Cope and Marsh had already been born. The first American dinosaur was found by paleontologist Joseph Leidy in 1858, and the fact that Othniel Charles Marsh and Edward Drinker Cope launched a veritable hunt for fossils a little later, with the hunters armed not only with shovels and chisels but also revolvers and rifles, can be largely attributed to their extraordinary personalities. Each of them wanted to be the first to discover the next new species of dinosaur, and so colleagues became archenemies and discovery sites were defended to the death. The two men ruthlessly covered up outcrops with dirt, destroyed bone fragments, and in some cases blew up entire sites just to hide their finds from other researchers.

Cope and Marsh first met in Berlin in the winter of 1863, where Marsh was studying anatomy. Cope was on a trip across Europe, and both escaped a draft and the horrors of the American Civil War by spending time abroad, even though they were of military age. Even then, it helped to be rich. While hundreds of thousands died on the battlefields back home, the two gentlemen were able to study and travel in peace in Germany. Cope's father is said to have sent his offspring to Europe primarily because of an affair that he thought was not befitting his son's station and thus wanted to stop. When Cope and Marsh first met, they were initially on good terms. They were even said to have taken walks through Berlin together and later wrote letters to each other, and they exchanged fossils, photographs, and manuscripts.

However, they were very different. Cope was the eager autodidact from a rich Quaker family with an astute academic mind, and the rather sedate but extremely methodical Marsh came from a poor background, but had a rich uncle who endowed him with enormous financial resources. This uncle was George Peabody, a businessman, investment banker, and one of the great philanthropists of his time. Thanks to his uncle, Marsh was able to attend Yale University, where he studied geology and paleontology, a relatively new subject at the time, and then anatomy in Germany. After the two met in Berlin, it didn't take long for tensions to grow and the first small intrigues to develop between them. This was probably due in part to the fact that Marsh had already earned two academic titles at the time but had published only a few scientific articles. Cope, on the other hand, who was almost nine years younger, could already boast thirty-seven scientific publications, even though he had dropped out of school after the age of sixteen.

These two men contributed a lot to the research of dinosaurs and other prehistoric animals. Cope was a brilliant mind, describing more than a thousand new vertebrate species and publishing some fourteen hundred scientific articles during his lifetime, a record that remains unbroken today. Marsh, who was in his final year of graduate school when Charles Darwin published his book on the origin of species, was one of the first scholars to adopt Darwin's theory of evolution in the United States. He was an early proponent of the view that birds were closely related to dinosaurs and theorized in 1877 that they were the descendants of these prehistoric animals. The reason that he came to this conclusion is probably linked to the fact that the prehistoric bird *Archaeopteryx* was first scientifically described in Germany in 1861—just a few years before Marsh's trip there—and his money certainly made it possible for him to see the fossil for himself. He was actually a geologist and not as well-versed as Cope when it came to vertebrate anatomy, but he had good connections in influential circles and was a shrewd strategist with political skills. As important as both men were to dinosaur research, they were apparently difficult contemporaries. Marsh was said to be consumed by ambition—overreaching, unscrupulous, and egotistical. Cope was described as quick-tempered and overbearing. Naturally, it didn't take long for the first friction to develop.

As soon as they returned to America, the two scientists were appointed to different universities: Marsh got the first chair of paleontology in the United States at Yale University, and Cope became a professor of zoology at Haverford College, but only after his parents pulled some strings. He was awarded an honorary master's degree, without which he could not have become a professor. Both scientists initially named a few new species after each other, which is a great honor in paleontology. Cope named a fossil amphibian *Ptyonius marshii*, and Marsh returned the favor by naming a giant marine reptile *Mosasaurus copeanus*. But Marsh's gesture was less friendly than it appeared. For as it turned out, the mosasaur came from a quarry in New Jersey where Cope had previously worked. He had shown this quarry to Marsh in 1868, and Marsh, behind Cope's back, had arranged with the landowner that new fossils from there would be shown to him first. Thus, the first description of fossil species quickly became a competition.

Cope considered Marsh intellectually inferior and claimed—initially behind his back, later publicly—that Marsh would have never been able to publish anything without his support. He also accused his colleague of having used him only to take advantage of Cope's intellect. It was all the more humiliating for Cope when Marsh, of all people, made fun of him in front of the whole world when, in reconstructing the marine reptile *Elasmosaurus*, Cope had confused the tail and neck of the animal and mistakenly placed the head at the tail end. Marsh used this gaffe to ridicule the "prodigy" Cope.

Cope's publication on *Elasmosaurus* appeared in 1868. After that, the two disputants never spoke favorably of each other again, and from 1873 on their arguments were downright hostile, accusing each other, among other things, of incompetence. The real bone war then started in 1877, when Marsh received a letter from a schoolteacher in Colorado. This teacher—an amateur fossil collector—and an acquaintance had been hiking in the mountains near the town of Morrison, where they were looking for fossilized leaves in the sandstone. Instead, the two found large bones embedded in the rocks. The teacher wrote in the letter to Marsh that the bones were probably a vertebra and a humerus of a giant lizard. He sent Marsh a total of about seventeen hundred pounds of fossilized bones. But at one point, when Marsh did not respond to a letter right away, the teacher also sent Cope some of the specimens. Marsh heard about this and published his research findings as soon as possible thereafter. He also paid the teacher a hundred dollars for the finds, which prompted the teacher to write to Cope again, asking him to forward his specimens to Marsh. This offended Cope immensely. However, in March 1877, Cope received bones from a school principal in Cañon City, also in Colorado. These remains were from a dinosaur that must have been even larger than the one Marsh had described. It was a long-necked dinosaur that was given the name *Amphicoelias fragillimus*. *Amphicoelias* is hardly known to us today, because unfortunately Cope's original material was destroyed or lost. Among experts, however, his species is considered a hot contender for the record for the largest dinosaur ever found. The animal has since been renamed *Maraapunisaurus*, because *Amphicoelias fragilimus* was only know from fragmentary material at the time, and while it was believed to be a diplodocid sauropod, more recent phylogenetic analyses show that it was in fact a primitive rebbachisaurid. Hence, the name had to be changed.

This incredible find of Marsh's rival prompted Marsh to act swiftly. When he learned of a huge bone bed northwest of Laramie in Como Bluff from two workers from the Union Pacific Railroad Company, which was building a transcontinental railroad across the States at the time, he sent one of his men there to take care of the excavation. Como Bluff turned out to be a gold mine for fossils; it was a veritable dinosaur graveyard. However, shortly after, Cope learned of the site as well and assigned his own team to search for bones there. In the process, each team sabotaged the other's progress as best they could. Marsh tried to use his money and political connections against Cope and to prevent his publications. Cope enticed some of Marsh's men to work for him instead. Because both feared that their workers might change camps and reveal discovery sites, they were extremely discreet about the exact locations of their fossils. When Henry Fairfield Osborn, then a student of paleontology at Princeton, visited Cope to ask him where fossils could be found in the American West, the latter politely refused to answer. Mr.

Osborn would still go on to play a big role in the discovery of another famous dinosaur, later, as we'll explore later in this book.

The most famous finds from the Bone Wars—and the most famous finds from the Morrison Formation overall—all came from the Como Bluff site. The excavation site is located in western Wyoming and contains a sequence of Upper Jurassic sediments. It was from there that the following dinosaurs were first described: *Allosaurus*, *Stegosaurus*, *Apatosaurus*, *Brontosaurus*, and *Diplodocus*.

The work at Como Bluff also marked the climax of the disputes between the two scientists, and the methods they used to try to outdo one other got completely out of hand. The damage done to science with these reckless actions is immeasurable. To defend himself against Marsh's coercion of his men, Cope even hired some gunslingers, and on one occasion the fossil hunters of both camps threw rocks at each other. It was pure luck that no one was hurt in the process. Even the legendary bison hunter William Cody, who went down in history as Buffalo Bill, worked for Marsh at one time, supplying him and his team with fresh bison meat.

The Bone Wars lasted over fifteen years, leading to a mess of names for the extinct giants—and two broken men who died impoverished and still at odds with each other near the end of the nineteenth century.

CHAPTER 4

Germany's Dinosaurs

How, on an Island, a Giant Turned into a Dwarf

The dinosaur books I read during my childhood were always about the same dinosaurs: *Tyrannosaurus, Triceratops, Stegosaurus, Brontosaurus,* and *Ankylosaurus.* We know all these animals from discoveries in the United States. Most of my books were by American authors and had been translated into German, so unsurprisingly they described the faunas of the Jurassic and the Cretaceous of North America. The spectacular finds from China had not yet been made at that time, and reports about dinosaurs from South America had not found their way into popular science books. As a child, I had already heard about the discoveries of the first dinosaurs in England more than one hundred and fifty years ago, and about the expeditions to Africa that took place around 1910. But one learned nothing about current research from these countries in German children's books, and very little about discoveries in Germany. The only exceptions were *Plateosaurus* from the Triassic, and *Compsognathus* and *Archaeopteryx* from the Jurassic. Because Germany was largely covered by the sea during the Mesozoic, and dinosaurs lived exclusively on land, we mainly find ichthyosaurs, marine crocodiles, crinoids, and ammonites. Both *Compsognathus* and *Archaeopteryx* come from lagoon deposits. They did not live there but were washed into these shallow marine areas. In such sediments we normally do not find any larger dinosaurs, and instead find fish, horseshoe crabs, and occasionally a pterosaur that tragically fell into the lagoon. However, because the Jurassic Sea was relatively shallow in large parts, there were islands in it, and *Europasaurus* comes from one of these islands.

PHOTO BY THE AUTHOR

A mounted skeleton of Europasaurus *at the* Mega 2015 *exhibition in Makuhari Messe, Japan.*

Europasaurus is the dinosaur to which I have the strongest personal connection, and I've researched it more than any other. For my diploma thesis, I examined the occiput and the inner ear of some long-necked dinosaurs with the help of computer tomography, one of which was *Europasaurus*. The original material of this animal is housed in the Dinosaur Park Müncheha-gen, where I later worked for a while. I lobbied on behalf of the park and sold a replica of a complete skeleton of *Europasaurus* to Japan, which became part of a special exhibition east of Tokyo. I got to travel to the opening, help set up the skeleton, give a talk, and was featured in a short welcome video. The show was called *Mega 2015*, and the booklet accompanying the exhibi-tion included a short article I wrote about *Europasaurus* and the area where it was found.

I have given more talks about *Europasaurus* than any other di-nosaur, and to the most diverse audiences. I spoke not only to exhibition visitors in Japan, but also to a large number of interna-tional colleagues. I even gave a talk as part of the Children's Uni-versity in Göttingen. That was particularly fun for me. The event was called "*Europasaurus*: A dwarf among giants!—Dinosaur re-searchers as detectives." The children were very engaged with the topic. They asked exciting questions about dinosaurs and fossils, and I showed them some original bones of *Europasaurus*, all of which were very small, although long-necked dinosaurs usually grew huge. Together we considered whether they might be the bones of a juvenile animal and concluded that they must instead be from an unusually small adult animal.

The explanation truly sounds like a detective story. *Euro-pasaurus holgeri* was first discovered in northern Germany. The genus name means "the lizard from Europe," and the species name honors the finder of the first remains, Holger Lüdtke, a fossil collector from the region. The fossil was discovered at the Langenberg quarry in the Harz Mountains, where sediments from the Upper Jurassic, about 154 million years old, are ex-posed. This quarry has turned out to be an exceptional fossil deposit. Its limestone and marl layers were tilted by about sixty to seventy degrees as a result of tectonic movements and now form a steep face. They have yielded an abundance of unique vertebrate fossils representing a specialized island fauna, from the Upper Jurassic, never described from anywhere else in the world. I have been to this quarry a few times to look for fossils there. Unfortunately, I did not find any dinosaur remains—but I did find a very large turtle. Its carapace was at least thirty inches in diameter, and I could not carry the heavy block in which the fossil was enclosed, not even with two colleagues. The boulder lay on a huge pile of other boulders that had been scattered across the hillside after a blast. Gravity helped us roll the boulder down the slope, and at the base of the rock face, a quarry employee loaded it onto a forklift and hauled it away. The turtle's carapace, which was clearly visible on the rock surface, had several bite marks. But it is difficult to say who caused the circular wounds. There were many large predatory marine reptiles in the Upper Jurassic, but the culprit was most likely a crocodile, as remains of at least four different crocodile species were discovered at the Langenberg quarry. In addition to crocodiles, turtles, and dinosaurs, small mammal teeth, flying reptiles and lizards have also been discovered there. However, the most remarkable find from the Langenberg quarry is actually the *Europasaurus*. While some genera of long-necked dinosaurs are among the largest land animals of all time, *Europasaurus* represents the first unequivocal dwarf sauropod and it is only known from this site. Histological studies by Martin Sander and his team have shown that some of those specimens were already fully grown despite their small body size. While close relatives of *Europasaurus* such as *Camarasaurus* and *Brachiosaurus* reached lengths of more than sixty-five feet, the maximum body size of *Europasaurus* was probably less than twenty-six feet, with a body weight of about eighteen hundred to twenty-six hundred pounds. In 2006, Martin Sander and his team concluded that *Europasaurus* was an example of a so-called island dwarf. During the animal's lifetime, a sauropod population was apparently cut off from the mainland and evolved into a dwarfed form in just a few generations under isolation on an island in the Lower Saxon Basin, while its closest relatives on the mainland remained giants.

> **Island dwarfing is an evolutionary biological phenomenon in which the size of animals living in an isolated**

ecosystem decreases within a few generations. Mostly this happens on islands, where these animals do not have to fear predation pressure. However, this phenomenon can also occur in other habitats such as caves, isolated valleys, or inaccessible mountainous regions. Other examples of island dwarfing include *Homo floresiensis*, a small-bodied relative of modern humans that lived on the Indonesian island of Flores until about sixty thousand years ago; the dwarf mammoth from the Russian Wrangel Island off the coast of Alaska, where such mammoths still existed until about thirty-seven hundred years ago; and the dwarf goat *Myotragus* from Mallorca, which became extinct about forty-five hundred years ago. Dwarf hippos have also been extinct in Madagascar for about two thousand years, although reports of individual sightings support the assumption that this species survived somewhat longer in remote regions of the island. An example of an extant dwarfed species today is the Sumatran rhinoceros of Indonesia. Island dwarfing can be caused by an evolutionary advance in sexual maturity or by a complete cessation of an animal's individual development. Both can result in a retention of juvenile characteristics throughout later life stages, i.e., paedomorphosis.

The high number of individuals at different growth stages at Langenberg provides a unique opportunity to study ontogenetic stages (developmental stages) and the interspecific variability of this small sauropod. Such an ontogenetic range cannot be found anywhere else in the world. The excellent preservation of a large number of skull elements from at least fifteen individuals and postcranial skeletal material from more than twenty animals has allowed scientists to conduct a detailed phylogenetic study of these *Brachiosaurus* relatives. It is an important prerequisite for understanding the evolutionary history of this sauropod group, because as the continents were in motion and drifting apart in the Upper Jurassic, dinosaur faunas evolved differently in the south and north. In the process, some European dinosaurs of this time were linked with contemporary North American faunas, while others appeared to be associated with Chinese faunas. Brachiosaurids, which include *Europasaurus*, were found on both the northern and southern continents.

From the abundance of skull bones, we chose the best-preserved brain case of *Europasaurus* and scanned it at the University of Bonn using a computer tomograph and then created a 3-D model of the brain, cranial nerves, and inner ear with modeling software. Since I had already had the opportunity to perform such scans myself during my time at Bonn, I also created these computer models.

In 2015, two colleagues and I presented our results at a meeting in Dallas, Texas. For our research, the dinosaur park had loaned us a few specimens in advance and neatly packed them in boxes that were liberally lined with foam. The fossils had been stored on the premises of the University of Bonn for quite a while, and when I later flew to Japan for the special exhibition, I had to deal with this very fragile and delicate material again. Some of the original bones of *Europasaurus* were to be part of the exhibition. Fortunately, the valuable specimens were never harmed, because I always handled them with extreme care. Until one day, unfortunately, a mishap happened: I was in a hurry and ran with a suitcase from my office to the computer tomograph, which was located in another building. As I rushed down the stairs, I missed the last step and fell—and the suitcase slipped out of my hand. It sailed through the air before my eyes while sheer horror gripped me. I had banged my knee pretty badly, but at that moment I felt no pain. The case contained the best-preserved brain case of a *Europasaurus*, and when I opened it, I saw that a piece had broken off the side. I was so startled that tears welled up in my eyes. This was a unique, especially valuable piece, and I had damaged it. I was distraught and immediately ran to Olaf, our preparator. He noticed that I was completely upset and calmed me down. Since the piece had detached with a clean break, Olaf had no trouble reattaching it with a special fossil glue.

The results ultimately showed that the examination of the skull had been worthwhile. Since its anterior part was incomplete, the olfactory bulbs and parts of the cerebrum could not be reconstructed, but we were able to accurately re-create the posterior region of the brain and most of the cranial nerves. In addition, we were able to represent various vascular structures, such as the lagena and inner ears, the animal's organs of balance.

At the Langenberg quarry, we found remains of *Europasaurus* only up to a certain stratigraphic horizon, and after that there were no more finds of it at all. For a long time, it was unclear why *Europasaurus* suddenly disappeared from the fossil record. In recent years, however, we seem to have solved the mystery. In younger strata, only a few feet above the strata where *Europasaurus* was found, large footprints appear in the fossil record of the Langenberg quarry. They are probably directly related to the extinction of *Europasaurus* and other island dwarfs found in the region. The layer in which these footprints commonly appear is about sixteen feet above the *Europasaurus* horizon. Stratigraphers who have studied this rock sequence estimate that this corresponds to about a thirty-five-thousand-year time difference. The current working hypothesis is that a temporary sea level depression occurred, creating a land bridge between Sauropod Island and the mainland. This allowed predatory dinosaurs, which must have been very large, to invade the island. It is not quite clear which animals they were, but the shape of the footprints and sporadic finds of claws indicate that they must have been close relatives of *Allosaurus*, which we know from the Morrison Formation. One possible candidate is the *Neovenator*. Judging from the footprints, the animals must have been twenty-three to twenty-six feet long. The invasion of these predatory dinosaurs caused a faunal upheaval on the island. The carnivores that previously lived on the island were rather small, and apparently there had been a natural balance between theropods and europasaurs. For the giant *Neovenator*, however, the miniaturized long-necks were easy prey. Europasaurs could not oppose it and died out within a few generations.

The First Feather That Changed
Everything: *Archaeopteryx*

In science, coincidences often play a central role, and one such coincidence was the discovery of the prehistoric bird *Archaeopteryx*, first scientifically described by Hermann von Meyer in September 1861. Von Meyer's paper was about a single feather of a prehistoric animal—only mentioning a complete skeleton in a sidenote—that came from the same strata and had characteristics of birds and reptiles. Literally, he wrote: "From our living birds it shows some deviation."

The text is remarkable because von Meyer wrote it less than two years after Darwin's book *On the Origin of Species* was published. In this book, Darwin cited numerous proofs for his theory that all animal and plant species must have descended from a common ancestor. While von Meyer did not elaborate on Darwin's book, he seemed to have already known and accepted the theory. Darwin called the underlying process for the gradual change in all animal and plant species evolution. He saw its driving force as competition among species for resources and habitats, with natural selection favoring those forms that are stronger or better adapted to their habitat. With the publication of this book, a broad public was confronted for the first time with thoughts on the origin of species that contradicted the biblical story of creation.

What made Darwin's theory so difficult to grasp at the time was the lack of connecting links in the fossil record. Opponents of his theory argued that one would have to find substantially more transitional forms from one to the other species, should the species have actually emerged from each other. The fact that now, with the discovery of the "Urfeder" (German for "prehistoric feather"), there was actual evidence for Darwin's theory of evolution, immediately after the publication of his book, was certainly helpful in making his case.

The feather and the first complete skeleton of *Archaeopteryx* from 1861 were followed by another spectacular find of the prehistoric bird in the 1870s. This fossil is on display at the Natural History Museum in Berlin and considered one of the finest specimens. The first *Archaeopteryx* skeleton is now in the Natural History Museum in London. A total of twelve prehistoric birds and one feather have been found so far. In the case of one of these animals, it is now clear that it is not an *Archaeopteryx*, but a small, feathered predatory dinosaur that was probably more closely related to the sickle-clawed raptors than to birds. The other specimens also show differences that justify subdivision into different species. That is why some of my colleagues now avoid calling these animals *Archaeopteryx*; they simply call them "Urvögel" (proto-birds).

PHOTO BY THE AUTHOR

The London specimen of Archaeopteryx *from the Solnhofen Limestone on display at the Natural History Museum in London, United Kingdom.*

In 1860, British biologist and comparative anatomist Thomas Henry Huxley proposed that birds were descendants of dinosaurs. In doing so, he pointed to skeletal similarities between dinosaurs, *Archaeopteryx*—which he called the "first bird"—and modern birds. Marsh came to the same conclusion in the United States, and his initial descriptions of early Mesozoic tooth-bearing birds such as *Ichthyornis* and *Hesperornis* supported this thesis.

This view persisted until the end of the nineteenth century, when Danish paleontologist Gerhard Heilmann rejected the theory in 1926 and wrote in his influential book *The Origin of Birds* that birds could not be dinosaurs because dinosaurs did not have a wishbone. In birds, the left and right clavicles are fused to form the V-shaped wishbone. After that, there was silence on the subject for almost forty years, and it was generally assumed that birds must have had crocodile-like ancestors, until American paleontologist John Ostrom discovered a small carnivorous dinosaur in 1964 that he named *Deinonychus*, which means "the terror claw." We know these dinosaurs as the sickle-clawed raptors from the *Jurassic Park* movies. Ostrom noticed the unmistakable similarity of the skeleton to that of modern birds. He then also examined the skeleton of *Archaeopteryx* in detail, which led him to claim that Huxley had been right and that birds had indeed evolved from dinosaurs. To honor Ostrom's contributions, a feathered predatory dinosaur from the Altmühl Valley was named *Ostromia* in 2017. *Ostromia* is closely related to *Deinonychus* (a dinosaur which Ostrom named). Ostrom concluded that birds were dinosaurs because they're so similar to extinct dinosaurs. Funnily enough, *Ostromia* had long been mistaken for a "bird" (*Archaeopteryx*) which underscores how similar birds and maniraptorans (the group to which *Ostromia* and *Deinonychus* belong) really are. Incidentally, this *Ostromia* is the supposed *Archaeopteryx* specimen that was actually a raptor—and its namesake is therefore extremely apt.

Ostrom's work and the subsequent publications of paleontologist Robert Bakker initiated a veritable renaissance in dinosaur research and led to a profound rethinking of almost all aspects of dinosaur biology. They created the foundation for our current image of dinosaurs and for the way these animals are portrayed in film and on television. And it all started with the discovery of a single *Archaeopteryx* feather.

However, the discovery of *Archaeopteryx* is not only historically important. Especially the last three specimens discovered have contributed to our better understanding of these animals today. Oliver Rauhut and his team from Munich University were also able to distinguish them from another small predatory dinosaur that was also discovered in the Plattenkalk (a very finely grained limestone). They dubbed it *Alcmonavis*, "the bird from the Altmühl River."

In the tenth specimen of *Archaeopteryx*, which is excellently preserved, structures on the skull that could not be observed before were recognizable for the first time, and the bones in its foot showed unusual features. In the eleventh skeleton, the feathers are particularly well preserved, and it is possible to clearly see the pennaceous feathers. The research team was able to show that the feathers of *Archaeopteryx* were already asymmetrically shaped, as in modern birds. What this means and how important this discovery is will be explored in a later chapter, when we take a closer look at the evolution of birds.

The latest fossil to date was discovered in 2010 in a quarry near the Köschinger Forest north of Ingolstadt. It comes from strata of the so-called Mörnsheim Formation and, at around 153 million years old, is the oldest find of the prehistoric bird to date. The fossil is about as old as the strata from the dwarf island of Langenberg. This makes my ears prick up, because it is quite conceivable that one day we might also discover remains of a proto-bird from the Langenberg quarry. Martin Sander predicted this several years ago, and I am curious to see if he will be proven right. Should such an animal actually appear one day, I would call it *Sanderopteryx predicta*—"the predicted Sander's feather."

The Adventures of Werner Janensch

I grew up in the very south of Germany on beautiful Lake Constance and only came to Berlin for the first time after the reunification of East- and West-Germany. At that time, the first thing I wanted to do, of course, was to visit the Museum für Naturkunde in East Berlin, especially to see the Berlin *Archaeopteryx* specimen and the *Brachiosaurus* skeleton. *Brachiosaurus*, or "arm lizard," was a large long-necked dinosaur whose arms were longer than its hind legs, quite unlike many other sauropods such as *Diplodocus, Brontosaurus*, and *Apatosaurus*. The skeleton in the museum's Great Dinosaur Hall consists of original bones. To date, it is the largest complete, mounted original skeleton of a dinosaur in the entire world. During its lifetime, the animal must have been some seventy-five feet long and its head towered about forty feet above the ground. *Brachiosaurus* was found by the paleontologist Werner Janensch. Janensch was an eminent vertebrate paleontologist, geologist, explorer, and one of the most important German dinosaur specialists of his time. As a young man, he traveled to Africa to search for dinosaurs. Strictly speaking, the skeleton—or perhaps I should say the skeletons— of the Berlin exhibition is comprised of not just one, but at least two dinosaurs of the same species. Because Janensch found several partial skeletons that complemented each other well, it was decided during the mounting of this reconstruction, to build a composite from bones of various animals in order to be able to show as complete a skeleton as possible.

At the end of the nineteenth century, Reich Chancellor Otto von Bismarck, following the example of England and France, finally agreed to acquire colonies in Africa after long urging by the imperialist movement in Germany. One of these colonies was called German East Africa. It existed from 1885 to 1918 and included the present-day countries of Tanzania, Burundi, as well as Rwanda and some areas of Mozambique. The colony had a total area of 384,000 square miles, almost three times the size of the Federal Republic today. German East Africa was the largest colony of the German Empire, with almost 8 million people living there by its end. After the end of World War I, in accordance with the provisions of the Treaty of Versailles, the colony was divided between Belgium and Great Britain and placed under the administration of the League of Nations.

The young Janensch found himself in German East Africa in April 1909. Three years earlier, a mining engineer there had discovered a huge bone sticking out of the ground on the slope of a hill. The locals called the hill Tendaguru, or the steep mountain. The Upper Jurassic rock sequence exposed there is called the Tendaguru Formation. It is considered one of the most important outcrops of Jurassic vertebrate fauna in the southern hemisphere. The Tendaguru mound is located near the port town of Lindi on the southeast coast of the former colony. Today, the town is part of Tanzania. The mining engineer who made the find informed his superior in Hanover, and he forwarded the report to the Commission for the Geographic Exploration of the Protected Areas. This department of the Imperial Colonial Office dealt with the geographic assessment of the German protectorates. Hans Meyer, a German Africa researcher, heard about the find through the grapevine, who in turn informed the paleontologist Professor Eberhard Fraas. Fraas was asked to examine the find on-site, and traveled to Lindi by steamship, arriving in August 1907.

After a five-day walk, Eberhard Fraas reached the Tendaguru mound. It was immediately clear to him that these must be dinosaur bones. Fraas also observed that the Tendaguru Formation was unusually rich in fossils. The mining engineer who made the discovery joined the professor and, with a local team, helped him recover two large skeletons of long-necked dinosaurs and ship them to Germany. These giants were later described as the genera *Tornieria* and *Janenschia*.

After his return to the German Reich, Fraas tried to raise enough money for an expedition to Africa. He found support from Professor Wilhelm von Branca, who was at the time the director of the Geological-Paleontological Institute and Museum of the Royal Friedrich Wilhelm University in Berlin. For von Branca, it was a matter of German national pride to help such a project succeed. He collected a large sum of money from his wealthy friends and sent his best man, curator Werner Janensch, to Africa. Janensch arrived in Dar es Salaam in April 1909 and led four field campaigns until 1912. However, because the region around Tendaguru was infested with tsetse flies, which transmitted nagana disease to horses and other pack animals, Janensch had to rely on the help of many local porters. In the first season, he employed about one hundred and sixty local workers. With their support, he opened about a hundred outcrops and shipped the recovered fossils to Germany.

In addition to the sauropods *Tornieria* and *Janenschia* discovered by Fraas, there were other long-necked dinosaurs in the Tendaguru Formation. One was the medium-sized *Dicraeosaurus*, which was relatively common in these strata and is now also on display in the Berlin Natural History Museum. But the find that made Janensch famous was a larger sauropod that came to light around half a mile northeast of the hill. Janensch made the discovery of the giant back in June 1909 and later named the animal *Brachiosaurus brancai*. The first find consisted of a few tail vertebrae, the sacrum, the pelvic bone, a few ribs, the upper and lower arms, shoulder blades, femora, and the calf and shin bones. Although this was a juvenile that was not quite fully grown, the humerus was already five feet three inches long and the femur five feet one inch. A second find followed in August 1909, when about one hundred and fifty bones of various dinosaurs were discovered in a large outcrop, including two femora of *Brachiosaurus*. In September of the same year, Janensch found more bones in a third pit, widely scattered, along with bone fragments that were already heavily weathered and could not be precisely identified. Finally, in October 1909, he made his most important find, south of the hill, on the bank of the Kitukituki River, where two skeletons of *Brachiosaurus* lay at once. At this place the bones were so abundant that Janensch continued to dig there until 1912. He assumed that the animals must have sunk into the mud because they were found in an upright posture with their arms and legs stuck vertically in the sediment.

In total, Janensch must have found at least seven individuals. How many there actually were is unclear. He did not keep accurate field notes and did not document all the sites where they were found. In addition, some fossils from Tanzania were destroyed by the Allies during an air raid on Berlin in 1943. Janensch had collected so much material that at that time not all the boxes from Africa had been opened and examined yet, which is probably why many a treasure was lost at that time. Janensch spent the rest of his career examining and describing the finds. As he worked on the *Brachiosaurus* material, he noticed that many elements bore a strong resemblance to an animal from the Morrison Formation that American paleontologist Elmer Riggs had described back in 1903. He had named it *Brachiosaurus altithorax*. Janensch chose the same genus name for the long-necked dinosaurs from the Tendaguru Formation, but added a new species name, *brancai*, to honor Professor von Branca.

When paleontologist Gregory S. Paul examined the skeletons of *Brachiosaurus altithorax* and *Brachiosaurus brancai* in 1988, he concluded that it was justified to establish a subgenus for both animals. Paul was aware of the historical significance of the names and suggested that the animals be called *Brachiosaurus (Brachiosaurus) altithorax* and *Brachiosaurus (Giraffatitan) brancai*. In practice, however, this designation with subgeneric names was never used. The situation did not change until Mike P. Taylor, another paleontologist, published an article in 2009 explaining that there were actually so many differences in these animals that they needed to be given different generic names. He named Janensch's animal *Giraffatitan brancai*. I remember the article very well because I had resumed my paleontology studies in Bonn shortly before it appeared. My first lecture in the seminar was a summary of the article, and we discussed whether this renaming was justified. Coincidentally, Mike Taylor was visiting Bonn a little later, and I was also able to talk to him personally about this issue. One of his arguments was that Janensch had probably never seen the original bones of *Brachiosaurus altithorax*. However, I wonder if this argument is relevant. I have seen both dinosaurs with my own eyes, and they seemed very similar. In addition, Taylor argued that the number of distinguishing features between the two animals was so great that they should definitely be separated by name. I find this argument understandable, though not tenable. Indeed, the International Code of Zoological Nomenclature, which lay down the naming and classification of all animal species internationally, do not specify the number of characteristics in which one animal must differ from another in order to belong to a different genus.

This is not trivial at all. To better explain, let me give an example. In today's animal kingdom there are the so-called true big cats, which carry the scientific name *Panthera*. The big cats include the tiger (*Panthera tigris*), the jaguar (*Panthera onca*), the lion (*Panthera leo*), the leopard (*Panthera pardus*) and the snow leopard (*Panthera uncia*). All of these animals look very different, live in different habitats, and some are of different sizes and weights. Nevertheless, they belong to one and the same genus. Nobody would think of splitting this genus. Just because the two long-necks differ is not a sufficient reason to divide them into two different genera.

Only if other dinosaur genera would stand between the two in the dinosaur family tree would one need to change the name.

But the only dinosaur that might qualify for this scenario is *Sonorasaurus thompsoni* from Arizona. In a 2016 study, *Sonorasaurus* is phylogenetically sandwiched between *"Giraffatitan"* and *Brachiosaurus*, but in another 2017 study, *"Giraffatitan"* and *Brachiosaurus* are more closely related, and *Sonorasaurus* is only a sister taxon to *"Giraffatitan."*

The problem with the 2016 phylogenetic tree would be that the genus name *Brachiosaurus* would be split and another genus would be stuck in between. According to the rules of zoological nomenclature, this is not possible. It would make the genus *Brachiosaurus* paraphyletic. This means that *Sonorasaurus* would be split off from a larger group, even though it has fewer differences, since it would lie between the other two, and the remaining remnant, i.e., *B. altithorax* and *B. brancai*, would remain lumped together in the absence of other visible features. This would indeed be wrong and would not make sense. But even if we follow the 2016 phylogenetic tree, there are two possibilities: either we actually rename *B. brancai*, or we would have to change *Sonorasaurus* to *Brachiosaurus thompsoni*. From my point of view, the latter would be more justifiable, since the name *Sonorasaurus thompsoni* has not been around that long. The exact classification will perhaps become more interesting when the brachiosaurid, found in Portugal in August 2022, is examined and scientifically described. It is already considered the largest dinosaur ever found in Europe. Which species it is, however, is still uncertain.

THE LOWER CRETACEOUS

(145 to 100.5 million years
before present)

Iguanodon

Polacanthus

Neovenator

Baryonyx

CHAPTER 5

Argentina—Where the Giants Live

The Long-Necked Dinosaurs That Tower over Everything

For many decades, the Berlin *Brachiosaurus* skeleton found by Werner Janensch was the largest complete original skeleton of a long-necked dinosaur in the world that could be admired in an exhibition. Other candidates for the title of largest land creature mostly came from the United States. But in the early 1990s, enormous bones of much larger giants were found in Argentina. These giants belong to a group of long-necked dinosaurs that are called titanosaurs, which survived into the Cretaceous period. The very first finds of titanosaurs from South America were made about one hundred and thirty years ago, but it was not until the 1990s that remains of long-necked dinosaurs were found that stretched the boundaries of what was deemed physiologically possible. With the description of *Argentinosaurus* in 1993, the world was introduced to a dinosaur that must have been almost 100 feet long and, depending on which author and which calculation method one may follow, probably weighed between sixty and ninety tons!

Sauropods are unique in many ways. We know of no comparable creatures in the animal kingdom today. They have a build that does not exist in any other group of animals. The obvious and most important feature is their neck. No other animal has such a long neck as the long-necked dinosaurs. Surely many will automatically think of a giraffe at this point and suggest that giraffes also have long necks. However, this is relative—the neck of an adult male giraffe can grow up to six feet long. This corresponds to the average height of a human male. From our point of view, that's enormous. But some dinosaurs had necks up to forty-five feet long. By comparison, the neck of a giraffe is downright tiny. To produce such an extremely long neck, the sauropods developed first one and then a second additional cervical vertebra and also shifted the shoulder girdle backwards, so that functionally, the first dorsal vertebra became another cervical vertebra. In addition, each cervical vertebra became significantly more elongated, while the weight of the vertebrae was drastically reduced. This was possible because the walls of the vertebrae were paper-thin and interspersed with a complex network of air sacs. These air sacs provided efficient respiration through unidirectional airflow, and also enabled the sauropods to better remove excess heat and avoid overheating. This was necessary because the long-necked dinosaurs, at least in the early stages of their lives, had an active metabolism and grew rapidly. The rapid body growth, the long neck, their avian lungs, and an efficient digestive tract ultimately enabled the sauropods to grow so large.

The dwarfed *Europasaurus* is a special case among the long-necked dinosaurs, because most sauropods were giants—true titans! No other land animal was even close to that size. The African elephant, the largest land creature today, is a dwarf compared to the "lizard-footed"—that's what the term "sauropod" means in English. The largest sauropods were ten times, perhaps even twenty times, the weight of an elephant. One has to imagine enormous colossi, some of which must have been as heavy as a jumbo jet or a sequoia tree. These animals were mountains of flesh, like long-necked blue whales on four legs. It is not without reason that the first sauropod ever described was also named *Cetiosaurus*, "the whale-lizard." I like this name very much, because it is so pictorial. In my opinion, the long-necked dinosaurs have the most fitting names.

One of the most beautiful names belongs to the first dinosaur I remember learning about as a child: *Brontosaurus*, "the thunder lizard." He is also a long-necked dinosaur. As a little boy, I always imagined how the Earth must have shaken with every step this giant took, and then I'd stomp through the house with a loud roar, much to the chagrin of my parents and siblings. Another giant was named *Seismosaurus*, "the earth lizard." Later it turned out that it was a particularly large *Diplodocus* species, hence the genus name has since become invalid. Basically, most sauropod names contain an allusion to their enormous size: *Seismosaurus*, *Cetiosaurus*, *Brontosaurus*, *Supersaurus*, *Ultrasauros*, *Notocolossus* ("the colossus of the south"), and, of course, *Titanosaurus*, namesake of the titanosaur clade. They get their name from the Titans of Greek mythology. These were giants in human form, a powerful race of gods. That paleontologists chose such names to describe the enormous size of sauropods is understandable, especially when you compare these creatures to extant animals. The smallest dinosaur of our time can illustrate this impressively: bee hummingbirds (*Mellisuga helenae*). These very beautiful birds live in Cuba, with the plumage of males and females iridescent in different colors. The females weigh no more than a penny (0.08 ounces) and the males less than 0.07 ounces, which is about the weight of a gummy bear. Nevertheless, they are true theropods; that is, they belong to the group of dinosaurs that also gave rise to *Tyrannosaurus*, the king of all carnivores.

The largest sauropods are now known to have weighed at least fifty tons, and many of my colleagues even consider body masses of over one hundred tons possible. This means that the heaviest dinosaurs weighed 50 million times more than the lightest members of this group! Fifty million times more! That is fifty thousand multiplied by one thousand. But how does one calculate the weight of a dinosaur? One of the methods for determining the body mass of four-legged dinosaurs involves the circumference of the thigh bone and the humerus. This is certainly easy to understand, because the extremities have to carry and move the weight of the animal around. The heavier the animal, the stronger the arm and leg bones must be to hold the corresponding muscle mass. Most researchers agree that the crown for the heaviest dinosaur currently still belongs to *Argentinosaurus*. *Argentinosaurus*, however, is known only by its thigh, a few vertebrae and a few small bone fragments. This femur is indeed massive, but I find it difficult to accept it alone as a reliable reference for calculating its mass. In general, most skeletons of the long-necked dinosaurs we know today are extremely incomplete. In 2014, however, a team of researchers discovered an entire herd of titanosaurs in Chubut Province, Argentina. I have a very special relationship with these giants because I know the paleontologist José Carballido, who recovered the animals and later scientifically described them, very well. He worked as a postdoctoral researcher in Bonn while I was studying for my degree. In 2014, when José discovered this new colossus, I had just started a job as a scientific advisor for Dinosaur Park Münchehagen. At the time, *stern TV* was reporting on the giant find, and I had been invited to appear on the show on behalf of the park. While we were talking about the very largest dinosaur ever found, I prepared a fossiliferous block live on air that contained bones from the shoulder girdle of the dwarf sauropod *Europasaurus*, the smallest representative of this group. Since the dinosaur park had a subsidiary that produced life-size dinosaur models, I was asked at that time to establish contact with the museum in Argentina, and after long negotiations, the world's largest live reconstruction of a dinosaur was finally built under my guidance and supervision. The model is truly gigantic. Originally, it was supposed to have been exhibited in an old warehouse in Trelew, Argentina. But since the animal was too big for the hall, the model now stands in the open, outside the city near Trelew International Airport.

Many more bones are preserved from this titanosaur than from *Argentinosaurus*, including the humerus and femur. To reconstruct the anatomy of the animal, bones from at least three individuals of similar size were examined. José then gave the dinosaur the scientific name *Patagotitan*, "the Titan from Patagonia." After applying various methods, a reliable mass estimate of about sixty-nine tons was obtained. However, the standard error in such a calculation is about seventeen tons. So the animal could actually have weighed between fifty-two and eighty-six tons! This sounds like a large margin of error, but it is relatively small compared to estimates for other sauropods. We have a great many skeletal elements of *Patagotitan* available and can use several different measurement methods, all of which confirm a similar weight for the titanosaur. For *Argentinosaurus*, such mass estimates cannot be calculated with scaling equations because of the lack of the humeri, which makes the standard error much larger for it. In addition, the anterior dorsal vertebràe of *Argentinosaurus* are about 10 percent smaller than those of *Patagotitan*. Thus, *Patagotitan* actually represents the largest known dinosaur species. Probably more finds of giants are to be expected from Chubut, and I would not be surprised if one day an animal is discovered there with a size that would put even *Argentinosaurus* and *Patagotitan* in the shade.

The museum in Trelew has made many casts of the bones of *Patagotitan*, and replicas of this giant are now located all over the world. In the Field Museum in Chicago, there is a model in the main hall that is about one hundred and twenty feet long and over twenty-eight feet high. In late February 2022, during a trip to Argentina, I visited the workshop in Trelew that makes these giants. At the time, I was to participate in a campaign by Professor Oliver Rauhut from Munich, who researches and was searching for dinosaurs from southern Argentina. This was at a time when international air travel was extremely restricted due to the Covid pandemic. Oliver didn't know if this campaign would even take place and couldn't confirm it until a week before my flight. Fortunately, my Covid test came back negative, and a short time later I was already on the plane from Frankfurt to Buenos Aires. After a one-day stopover, I continued to Trelew the next morning, where Oliver kindly picked me up from the airport.

Even at the Trelew airport, you feel as if you've landed in dinosaur country. On the wall of the baggage claim area, a relief of a dinosaur excavation site with bones of the prehistoric giants is emblazoned, and a little farther on there are posters of dinosaurs, advertising the local paleontological museum. It is a major attraction of this city and brings in many international visitors. Its exhibition is worth seeing, but the real treasures are in the collection, which is not open to the public. Among other things, the original bones of *Patagotitan* are kept there. One night I stayed at the Touring Club Hotel. The name sounds like it's a flophouse for backpackers, but it's actually a legitimate hotel with over one hundred and twenty years of history. Supposedly, the famous outlaws Butch Cassidy and the Sundance Kid stayed here for some time after fleeing the United States.

The next morning, we loaded up the two SUVs and headed west. We drove about four hours on National Route 25 to the village of Paso de Indios. Along the way, we saw many impressive Cretaceous and Jurassic rock formations, and the landscape often reminded me of the American Midwest. In Paso de Indios, we refueled and bought some provisions before continuing north. Route 12, one of the main north-south connections in Chubut, is an unpaved dirt road, and I soon realized why Oliver had chosen the off-road vehicles. His goal for this campaign was to investigate an older outcrop where parts of a long-necked dinosaur skeleton had come to light some time ago. He wanted to find out how much of the skeleton was still hidden in the rock so that he could then decide whether to recover it later with heavy machinery. Another focus of this campaign was to find new outcrops. Oliver has been working in this part of the world for over twenty years, but it was my first trip to Argentina and South America. I was completely overwhelmed by the beauty of its environment and the many animals and plants that were foreign to me. Most impressive to me were certain large insects, grasshopper relatives, which do not exist here in Europe, and the many birds of prey with their remarkable wingspans. But at the sight of the nandus and guanacos I fully realized that I was in an absolutely strange world. Except for in Kenya, I had never seen such large flightless birds in the wild as those, simply running through the wilderness here. These nandus sprint seemingly effortlessly, at up to thirty-seven miles per hour across the rocky terrain. The guanacos are a wild species of the llama genus, though the llama we know is a domesticated species. You just don't get to see animals like this in the wild in Europe.

We continued on Route 12 north along the Rio Chubut until we reached a small village called Cerro Cóndor. There were only a few individual houses, as well as a school that was apparently attended by the children of all the surrounding farms. Without the school, the village would probably no longer exist. Next to it was a small museum with some dinosaur finds from the surrounding area. Directly in front of the village was an outcrop of pterosaurs, which one of Oliver's doctoral students wanted to study more closely. A few miles farther north, about half an hour's drive away, we stopped to look for new dinosaur sites. The rocks there are assigned to the Cañadón Calcáreo Formation and are between 160 and 150 million years old. Thus, they range from the Oxfordian to the Tithonian, the oldest and youngest stages of the Upper Jurassic, respectively. Even if the giant titanosaurs of Argentina do not come from these strata, the rock is immensely exciting for us. Unlike the Upper Jurassic in North America, Europe, and Asia, we know little about this time in the southern hemisphere. Since the contest between Cope and Marsh, many North American Upper Jurassic sites have become known and been well studied, but from the southern hemisphere we only know the sites in Tanzania. Oliver has made it his mission to better study this period of Earth's history, and this is important because we can already see that long-necked dinosaurs evolved quite differently in the northern hemisphere than in Gondwana, the southern hemisphere continent. Furthermore, in no other country on Earth can we trace sauropod evolution in the fossil record for as long as in Argentina. Some colleagues even suggest that the cradle of sauropods was in Argentina. In fact, the oldest long-necked dinosaurs are already found in strata from the Norian of Argentina, a stage of the Upper Triassic that is about 228 to 208.5 million years old. Sauropods are still missing completely in the fossil record of North America and Europe from this time. But already in the Jurassic, these giants existed everywhere on Earth, and while their history was largely uniform into the Upper Jurassic, evolution in the Cretaceous progressed completely differently in the northern and southern hemispheres. In the United States and Canada, where long-necked dinosaurs were long the dominant group of herbivores, they became rare at the beginning of the Cretaceous until they then disappeared completely from Colorado, Wyoming, Montana, and North and South Dakota.

Only in the far south, in the region around southern Texas and New Mexico, small isolated colonies of long-necked dinosaurs remained in some cases. This was a species of titanosaur called *Alamosaurus*, while the long-necked dinosaurs in the north were successively replaced by the duckbill dinosaurs and the horned dinosaurs.

The situation was quite different in South America, where these ornithischian dinosaurs were almost absent. But even here, the *Diplodocus* and *Brachiosaurus* relatives were gradually replaced by titanosaurs, which attained an amazing species diversity. There were tiny titanosaurs—which sounds like an oxymoron—medium-sized forms, and huge giants; there were long-necked and short-necked forms, grazing animals and those that ate foliage from the tallest trees. Sauropods occupied a wide variety of ecological niches and thus prevented the development of other species. But in the Jurassic Cañadón Calcáreo Formation, north of Cerro Cóndor, titanosaurs were not yet present, and we therefore hoped to find other earlier forms of long-necked dinosaurs there.

PHOTO BY THE AUTHOR

On our quest for Jurassic dinosaurs, we drove through stunning landscapes in Chubut Province, Argentina.

It's always a pleasure for me to be part of a field trip and to excavate fossils that have lain dormant in the sediment for millions of years—but it's even more impressive when you discover your own outcrop with dinosaur remains. Sure, I helped dig at the pterosaur site, and I even helped dig up some long-necked dinosaur bones; but those outcrops had been discovered by others. I wanted to make my own discovery, to find a treasure myself and be the first to excavate fossil bone fragments in the middle of nowhere that could potentially deepen our understanding of the lives of those prehistoric giants. However, the search for these remains is not always crowned with success, even in an area as rich in fossils as Chubut Province. The area is simply very vast and you could sometimes wander around for days without finding anything. This can be frustrating, and that's what happened to me on this trip—that is, at least, until the last day! We had driven north in our SUVs and parked the vehicles near the river. Around ten o'clock, each of us headed off in a different direction; we had arranged to meet back at the car at 5:00 p.m. I hiked over hills and through small canyons, climbing steep slopes, always keeping my eyes on the ground in front of me. I turned over every stone that even remotely looked like a bone or shimmered bluish, because in this area the fossilized bones often have a pale blue coloring. Unfortunately, bluish lichens also grew on the stones here, so I often reached out in vain. In the afternoon, after having found nothing for hours, I finally stumbled upon a piece of eggshell from a nandu. It was lying on the scree slope of a larger hill, and I wondered if it might have come from an abandoned nest from up on the peak of the hill. Perhaps I could find another larger eggshell there to bring back as a souvenir for my children. I climbed up, but I couldn't find a nest anywhere. Instead, out of the corner of my eye, I saw a small bluish stone. Instinctively I picked it up, although I assumed that it was just another stone covered with lichen. But instead it was actually a piece of a fossilized bone. I wiped the dust off the surface and did what any paleontologist does when he wants to make sure a fragment is a bone in the field: I held the fragment's fractured surface to the tip of my tongue and licked it carefully. The interior of fossil bone consists of very porous material, and if you touch this surface with its tiny pores with your wet tongue, a slight vacuum is created and the bone sticks to your tongue. And lo and behold, the stone stuck. It was indeed a dinosaur bone! I was thrilled. I immediately set about searching the surrounding area for more pieces. I found a second one and then several more. In the end, I had five elements that all seemed to belong to the same bone. It was too big to be a turtle bone and too thick and too long to be that of a crocodile, but at the same time much too small to be the bone of a long-necked dinosaur. The only animal that made sense for it to have belonged to, in my opinion, was a carnivore. Judging by the size of the bone, the dinosaur seemed to have been quite large. I suspected that the fragments were parts of the fibula, because they appeared too thin for a tibia. I was pleased with this unusual find, since such carnivore remains are much rarer than those of herbivores. Since they appeared at the top of a hill, they could not have been washed there by the rain or any other water movement, and must have weathered out of the rock in place. When fossils are not transported by wind or water after they are deposited, they are called "in situ" finds. As excited as I was, I unfortunately didn't have a shovel with me to dig deeper into the rock. I had no choice but to take the exact GPS coordinates, photograph the site, and then with a heavy heart hike back to the off-road vehicles with the fragments.

I proudly showed Oliver my find in the evening and he confirmed that it was indeed the bone of a larger predatory dinosaur. Oliver had described a carnivore from the same sediments in 2017, which he had named *Pandoravenator*. My fossil was of similar size and was probably another specimen of *Pandoravenator*, and perhaps even a new species. Unfortunately, we could not identify it with certainty because we had to leave the next day and end the campaign. Another excavation with the appropriate equipment will hopefully reveal more material and then perhaps tell us what species it belongs to.

CHAPTER 6

Myanmar—Trapped in "Liquid Gold"

Feathers, Claws, and Ticks in Amber

Michael Crichton's cleverest idea in his 1990 novel *Jurassic Park* was to clone dinosaurs using DNA retrieved from their blood, found inside of mosquitoes. Crichton wrote that mosquitoes drank the blood of dinosaurs and immediately became encased in tree sap, which turned to amber over millions of years. Researchers then extracted the dinosaur blood, decoded the DNA, and cloned the dinosaur. This approach is currently impossible, yet the idea fascinates me to this day. There are fantastic amber deposits from the time of the dinosaurs, and some of them contain inclusions of spiders, snails, millipedes, and insects, some of which are actually mosquitoes. The oldest mosquitoes date back to the Cretaceous, but we have yet to see them from the Jurassic or Triassic. Consequently, a park of cloned dinosaurs could have included animals like *Tyrannosaurus*, *Triceratops*, raptors, and pterosaurs like *Pteranodon*, but not *Brachiosaurus* and *Stegosaurus*. Why the novel was nevertheless called *Jurassic Park* and not *Cretaceous Park* was probably due to the euphony of the name rather than the fact that the dinosaurs were really mostly

Jurassic. In fact, most of the dinosaurs in the book and the first movie came from the Cretaceous period. Furthermore, in the Jurassic World trilogy, which hit theaters as a sequel to the Jurassic Park film series, even marine reptiles like the *Mosasaurus* made an appearance. Now, that doesn't make any sense at all—how would a mosquito have been able to suck blood from this sea creature, and how could amber have been formed in the sea?

The most important amber deposits with inclusions from the Cretaceous period come from Canada and Myanmar. The fossils hidden in the amber of Canada are about 75 million years old and those from Myanmar about 99 million years. Because the country of Myanmar used to be called Burma, it is still often referred to as Burmese amber or Burmite.

Crichton's idea of extracting blood from mosquitoes was ingenious, but he could not have known that about thirty years later the first tiny dinosaurs would be found in the amber. These finds are truly breathtaking! Burmese amber gives us detailed information about what the plumage, soft tissues, and bone structure of juvenile birdlike dinosaurs were like, which we would not find under any other conditions.

What also makes amber fossils so special is their three-dimensional preservation. In the Solnhofen Limestone or in shale deposits, animals are usually flattened, so their actual anatomy is distorted. However, amber pieces are often small and rarely contain complete vertebrates, but usually only wingtips or the tail end of very small creatures.

One inclusion described in 2017 contained a very well-preserved hatchling of a birdlike dinosaur belonging to the enantiornithine (Enantiornithes) group. This group of feathered dinosaurs is very closely related to birds, even though it is outside the group of modern birds. Their strange name means "the opposite birds" and refers to the fact that these animals were the closest relatives of modern birds and resembled them in many aspects. However, these enantiornithines are now extinct.

The amber fossil in question holds a just-hatched juvenile that was only a few hours to a few days old at most. This is indicated by measurements of the skull, the wing, and the foot, which were compared with other juvenile and adult enantiornithines. The proportions of the fossil resemble those of an embryo rather than juveniles. The skull is larger, and the flight feathers are smaller. In addition, the small size suggests that it is a neonate. Because of the unique soft tissue preservation, for the first time, it was possible to observe in a dinosaur the external opening of the ear, the eyelids and scales on the legs, and the long talons on the toes, with one of the toes facing the others, so that the animal was probably able to hold on to branches well. The specimen also offers glimpses of the hatchling's down feathers, as well as the arrangement of its pennaceous feathers and even the color pattern of the plumage. This plumage has characteristics of both precociality and altriciality at the same time, which is not found in any modern bird. Thus, in the fossil we find strong flight feathers, while the rest of the body is sparsely feathered. The feathers on the legs, feet, and tail resemble the contour feathers of modern birds rather than their down feathers. We also find additional filamentous, or hairlike, feathers in the same places, which are comparable to the protofeathers of more primitive theropods. Overall, these amber finds provide us with a level of detail that fundamentally improves our understanding of the anatomy of the young of these pre-modern birds. In the process, it shows that the plumage of enantiornithines differs markedly from that of modern birds.

Although the above-mentioned animal was a hatchling, it apparently already possessed rectrices (flight feathers at the end of the tail, designed for steering), suggesting that it was capable of flight immediately after hatching. This is consistent with observations in other finds of enantiornithine juveniles, which are also thought to have been precocial (the young of precocial birds are already mature and mobile from the moment of hatching). Extant birds have a wide developmental range in their offspring, from helpless nestlings born blind and naked to extremely precocious fledglings able to fly as early as one day after hatching. In contrast, the range appears to be much narrower among enantiornithines, with every indication that their chicks were basically precocious. This certainly has to do with the limited variability in their egg size—all eggs of enantiornithines are about the same size. Limited developmental variability would also likely have prevented enantiornithines from breeding in some of the extreme environments used by extant birds today, where chicks would otherwise perish without intensive parental care. One example of adaptation to an extreme ecosystem is that of penguins, whose chicks cannot survive in the perpetual ice without their parents; another is the adaptation of birds to live on or in the water, where they must be either good swimmers or good gliders. So far, there is no evidence for either adaptation in the enantiornithines. We now know hundreds of aquatic birds, but so far, no enantiornithines with aquatic adaptations have been discovered. Very broadly speaking, it can be said that eggs of birds differ drastically in shape, color, and size depending on whether they breed in trees, cliffs, brushes, or in the sand. The degree of parental care in enantiornithines is not well understood today, but apparently their chicks were all highly precocial and arboreal at the same time. That is a combination we do not observe in any birds living today. All of today's tree-dwelling birds have chicks that are altricial. This means that the enantiornithines followed a developmental strategy that does not occur in modern birds, as the relatively early sexual maturity in the young, their slow growth rate, and their slow adolescence are atypical of birds today. There is a good reason for this: namely,

the retardation in growth meant that these young enantiornithines were longer exposed to the danger of predation, which
increased their mortality rate—especially if the parents could not
protect their chicks because they had left the nest early. This is
why we find so many juveniles of enantiornithines in the fossil
record and virtually none of modern birds.

The chicks from the Myanmar amber allow us to draw conclusions about the feather development of these avian relatives.
When we find certain types of feathers that are no longer found
in adults, or when we see feathers in adults that juveniles do not
yet have, we can see which feathers emerged only in the course
of their ontogeny and how they grew. This is especially relevant with feathers that we no longer find in modern birds. We
often see such unusual feathers on the legs and the tail. There,
a small number of protofeathers have been preserved, which
were part of the down plumage and were lost in the course of
evolution. The legs, the tail, and the belly of this enantiornithine were sparsely covered with such down, showing that it
had hatched only recently. His molt thus followed the pattern
of modern birds.

Not only are the feathers themselves exciting, but also the
ticks that are sometimes found on them. In 2017, such animals
were found on the feather of a dinosaur. Ticks are now among
the most widespread blood-feeding parasites that live on the skin
of a host. For a long time, nothing was known about the feeding
habits and possible hosts of Cretaceous ticks. The feather found
in 2017 now provided direct evidence that ticks fed on the blood
of feathered dinosaurs 99 million years ago. Unfortunately, what
kind of dinosaur it was is still unclear. Because we only have one
isolated feather, the host could have been an enantiornithine or
even a bird. Sickle-clawed raptors or egg thieves are also possibilities. So, in *Jurassic Park*, John Hammond, the director of the

dinosaur park, could have been looking for not only mosquitoes but also ticks in amber from which to clone his dinosaurs.

A Dinosaur Even Smaller than a Hummingbird?

In the competition for the most newly described dinosaurs, Othniel Charles Marsh and Edward Drinker Cope left behind a hopeless mess of invalid names and duplicate descriptions of these animals. This had to do with, among other things, the extreme ambition and egocentricity of the two paleontologists. Each wanted to outdo the other. Yet, both having been financially independent, they would have had enough time to examine specimens more thoroughly and diligently, to proceed with care and conscientiousness. After all, the number of scientific publications was low at that time, and the two were able to publish their reports without the need for lengthy proofreading. Both scientists were wealthy and not necessarily dependent on research funds—at least not at the beginning of their careers. Consequently, apart from their own ambition, there was no reason for them to rush out results. Today, things are different. Research moves so fast that hundreds of articles are published and up to fifty new dinosaur species are scientifically described every year.

Scientific articles must be proofread by at least two independent editors and checked for accuracy of content and plausibility. Only when the reviewers approve the article may it be printed in a scientific journal. This process is called peer review.

At the same time, scientists today are competing for research funding. Not every researcher receives sufficient funding to advance their field and publish their findings. There is often not enough money for new equipment, new computers, to attend meetings, or go into the field. A research proposal must be written for each grant, and the rejection rate is often high. This is especially frustrating for young academics. Those who want to stay in research must show universities that they are intellectually capable of writing research proposals and performing at a high level. Among the most important publications for paleontologists are the journals *Science* and *Nature*. For example, to publish an article in *Nature*, the topic must be groundbreaking, and the research must be pioneering and involve new methods. It must appeal to people beyond specialist circles.

The fact that what I perceive as one of the biggest blunders in paleontology in recent years happened in the renowned journal *Nature*, of all places, is extremely surprising. In 2020, a *Nature* article was published that had everything you would expect from a paper in this journal. Researchers had shortly before described a tiny reptile encased in amber. The find was a sensation, as it was said to be an exceptionally well-preserved birdlike skull. The skull appeared to belong to the smallest known dinosaur of the Mesozoic era, which probably weighed only a few grams during its lifetime, and competed with the bee hummingbird for the place of smallest dinosaur of all time. However, there was a catch to the story. The fact that spectacular articles have to be published ever more quickly today may also have led to a hasty publication in this case. What exactly happened and how did this mistake come about?

The researchers found a lump of amber containing the head of a small animal, and the rest of the body was missing. This head had large eye sockets, a round, domed skullcap, and a long pointed beak. When I saw photos of this fossil, I too immediately thought it was a bird.

Prior to this animal, only seven specimens of enantiornithine birds had been scientifically described from amber, which included several articulated skeletal elements. Six of these were juveniles and one was an almost fully grown adult specimen.

They were all smaller than the fossils that came from typical sedimentary deposits. But this animal was even smaller. Because it was also thought to be a bird with large eyes and jaws that still bore teeth, the researchers named it *Oculudentavis*, which means "eye-toothed bird." The animal was fascinating. In addition to dwarfing and avian features, it had a unique eye morphology that resembled that of lizards. In its eye socket was a ring of individual bones called a sclerotic ring, which indicated that the animal's pupil had originally been very small, although the eye socket, and thus probably the eyeball, had been very large. This bony sclerotic ring apparently indicated a reinforcement of the eye. It occurs in birds, in the extinct dinosaurs, pterosaurs, and ichthyosaurs, and also in some modern reptiles. The way these ossicles are arranged is more reminiscent of squamates than birds or dinosaurs. In every known dinosaur, the sclerotic ossicles are square shaped and narrow, and they usually delimit a larger opening inside the ring. In *Oculudentavis*, the sclerotic ring was very large, leaving only room for a small pupil. It was formed by elongated, spoon-shaped ossicles, otherwise found only in lizards.

Deep-diving ichthyosaurs had the largest eyes of any vertebrate and particularly pronounced sclerotic rings to withstand the enormous water pressure at the deepest depths of the sea. In the small pupils of *Oculudentavis*, my colleagues see evidence of a diurnal lifestyle. Its tiny body size suggests island dwarfing—according to them, the amber forest therefore once grew on an island. The unusual body shape of the animal suggests that this species occupied a previously unknown ecological niche. With the help of such amber deposits, we are able to uncover habitats that were inhabited by vertebrate species at the lowest limit of the body size spectrum. From the tip of the beak to the occiput, the head of this presumed dinosaur measures only 7.1 millimeters. Except for hummingbirds, there is no other group of birds that have such a small skull. The skull of *Oculudentavis* showed a general avian morphology with a slender beak that narrows at the tip, nasal openings located far back on its beak, enlarged and well-defined eye sockets, a short region behind the eye, and a dome-shaped skull roof. What should have puzzled colleagues, however, were the unusual teeth. They are stuck in the jaw in a way usually seen only in iguanas, monitor lizards, and geckos, and not in dinosaurs. Paleontologists assumed that this unusual feature was related to dwarfism, although no other detail, such as a shortened beak or toothlessness, indicated paedomorphosis. *Oculudentavis* had an entire row of small teeth. Therefore, the researchers believed that the dwarfing had been accomplished simply by reducing the growth rate. The way in which the individual bones of the skull were fused together and the shape of the sutures on the contact surfaces of the bones also did not conform to the pattern of theropod dinosaurs, which only emphasized the enigmatic nature of *Oculudentavis*. The researchers also tried to attribute this to structural constraints due to dwarfism. In the end, however, the differences in cranial anatomy were too great to be explained by processes of insular dwarfism. This was also noticed by many readers after the article appeared in *Nature*, which eventually led the authors to retract it.

Then, in August 2021, a new article on the subject appeared, clearly showing that the supposed dinosaur was in fact not a bird at all, but a small lizard. The large eyes and the strongly domed occiput had initially led the researchers down the wrong track. But a more thorough examination showed that several features disqualified the animal as a dinosaur despite its birdlike appearance. Another nearly complete fossil of this prehistoric animal showed that *Oculudentavis* was a squamate. The group of squamates includes nearly all extant reptiles today: lizards, iguanas, monitor lizards, and snakes. The only modern reptiles that do not belong to squamates are the turtles, the crocodiles, and the tuatara. An accurate classification in the animal kingdom had proven so difficult because *Oculudentavis* had apparently under-

gone a convergent evolution of its skull proportions that made it
look like a bird. However, other body features showed a clear affiliation with the squamates. When skeletal features are weighted
and combined with features from the gene database of extant
lizards, *Oculudentavis* appears to be closely related to an armless
and legless group of Southeast Asian blind skinks (Dibamidae).
If, however, skeletal features are weighted less, these odd animals
end up closer to tuatara in the phylogenetic tree. If only skeletal
features are considered, *Oculudentavis* lies somewhere between
iguanas and a group of extinct marine reptiles known as mosasaurs, which could grow up to fifty-nine feet long. With these
animals, *Oculudentavis* shares a long process of the premaxillary
bone that extends to the nose and the far-recessed nasal openings, but it differs from them in most other features of the skull.
It is all very complex. Despite this, or perhaps because of it, this
fossil is so extremely important to paleontology. *Oculudentavis*
was part of a forest ecosystem that is otherwise very rarely preserved in the fossil record, because forests are full of scavengers,
and the forest floor is just teeming with insects and worms clean-
ing up the mess. What is not eaten by scavengers and worms is
then taken by fungi. When an animal dies in the forest, there
is rarely anything left of it. The forest recycles very thoroughly
and is therefore a bad place for fossil preservation. The only way
for such small organisms to withstand the millions of years and
be preserved as fossils is to be sealed airtight in tree sap. Amber
is therefore a window to the Earth's past, giving us glimpses of
a fauna that would never have been preserved in sandstone or
mudstone. The conditions of deposition thus dictate what we
find later in the fossil record. In the fossilized tree resin, it is
small animals, plants, and pollen, whereas in the sedimentary
rock, it is large bones and tree trunks.

Fortunately, the amber deposits from Myanmar provide us
with remarkable details about the smallest vertebrates of the
Mesozoic and help us in their classification. For example, the

very next specimen of *Oculudentavis* discovered showed us that although bird features were present in the skull of the animal, there was no convergence with birds in the rest of the skeleton. On closer inspection, the resemblance to a bird skull could be attributed to deformation of the bones during fossilization, but because the rest of the body was missing from the first specimen, this misinterpretation was not immediately apparent. We now understand the ecology of *Oculudentavis* much better. Its small pupils indeed suggest that the animal was diurnal, and its embedding in amber suggests that it lived in trees. The long narrow mandible, the sharp pointed teeth, and the mandibular joint with little room for jaw muscles and an awkward lever indicate a weak bite force and a snapping mode of feeding for the animal, which apparently preyed on insects. The large eyes helped it to better identify its small prey. The fact that the relationships of *Oculudentavis* could be clarified was related to the fact that a second find contained parts of its body that helped identify the animal as a squamate. In addition, with the help of a computer program, the researchers were able to virtually recreate the original morphology of its deformed bones.

> **When bones are compressed, crushed or shifted against each other during the fossilization process, this is called "deformation." When a computer reverses this process in a simulation, it is called "retrodeformation."**

Retrodeformation restored the original appearance of the bones, which made them easier to compare with modern birds and lizards. The retrodeformation method supports traditional osteological studies, which will also help in the description of other fossils in the future. New finds from the amber of Myanmar will hopefully provide more specimens of *Oculudentavis*. In particular, finds with bones of its pelvic region, its tail, and its hands and feet would be important to determine an accurate phylogenetic position of these bizarre animals.

However, it is not only the discovery of new fossils in Burmese amber that poses a challenge to scientists; the extremely tense political situation in the country is also of great concern to my colleagues.

A Moral Dilemma

Fossils from the amber of Myanmar are spectacular because they preserve structures that usually have a low preservation potential and cannot be found anywhere else in the world. We can learn a lot about extinct species from this, because no fossil embedded in a rock matrix provides us with a comparably detailed picture of the world 99 million years ago. It becomes increasingly clear that the fossils of the faunal assemblage originated from an island and contained different species than other fossil deposits from the mainland. Perhaps someday it will indeed be possible to extract organic molecules, proteins, or even DNA from dinosaurs in amber, be it through blood ingested by trapped mosquitoes or directly through the cells, blood vessels, proteins, or other tissues of small dinosaurs. This prospect is tantalizing and promises groundbreaking insights in the future.

The amber from Myanmar, however, always plunges paleontologists into a moral dilemma. Myanmar is one of the poorest countries in the world and has been politically unstable for decades. Since the military coup on February 1, 2021, the situation there has deteriorated considerably. However, as the export of amber strengthens the illegitimate regime, many of my colleagues are discussing the reconcilability of scientific research with human rights violations. The Society of Vertebrate Paleontology, or SVP, a nonprofit scientific organization and the largest association of researchers, students, and interested laypersons in my field of research, has contacted its more than two thousand members several times to call attention to this human rights violation. The society, whose goal is to advance the science of vertebrate paleontology and to promote the discovery, preservation, and protection of vertebrate fossils and fossil deposits, first proposed a publication stop on all scientific articles dealing with fossils in Myanmar amber recovered after January 2021.

However, the troubling political developments in Myanmar came to a head when the country's armed forces seized political control in February 2021 through said military coup and declared a state of emergency. In the process, over seven hundred people were killed in just two months. The military had already gained control of the amber mines since 2017, repeatedly instigated armed conflicts, and carried out ethnic cleansings. The United Nations condemned these actions as genocide and crimes against humanity. Since the gemstone mines and amber deposits are located in the north of the country, where it is very difficult for international observers to reach, it was easier for the military government to commit atrocities there unobserved by media representatives and unpunished by the world. They secured access to the valuable mines and acted with great brutality against the local rural population. It was important to the SVP that its members and other paleontologists take special care in their research on the amber material from Myanmar. They wanted to make sure that fossils did not come from holdings of the military dictatorship so as not to indirectly support those in power in the country through research funds. The society admonished all scientists and scholars to refrain from publishing manuscripts on Myanmar amber acquired after the 2021 coup. However, the proportion of amber containing important vertebrate fossils is vanishingly small, while the majority of amber is processed in the jewelry industry. Nevertheless, it was important to take a stand on this issue. Even if one's own contribution does not necessarily lead to a change in the situation, it is still important to condemn the injustice and not be complicit in it. The fossils from this area of conflict are incredibly significant for our understanding of the evolution of the smallest vertebrates from the Cretaceous. We can only hope that peace will return to this beautiful country and that new spectacular fossil finds can be made in the amber that will reveal to us further secrets from a sunken world.

CHAPTER 7

The Spinosaurs

The Adventures of Ernst Stromer in Egypt

While Werner Janensch was uncovering giant long-necked dinosaurs in German East Africa, another German paleontologist and important dinosaur researcher was on the move in North Africa: Karl Heinrich Ernst Freiherr Stromer von Reichenbach. Stromer, who came from a wealthy noble family in Nuremberg, began his studies in geology and paleontology at the University of Munich in 1893, and wrote his dissertation on the geology of the German colonies in Africa as a doctoral student of Karl Alfred von Zittel. Africa fascinated Stromer throughout his life, and he came to Egypt for the first time after his habilitation in the winter of 1901. His first trip took him to El Fayum, a fossil site discovered by German botanist Georg August Schweinfurth, which yielded mammals from the Eocene and Oligocene. Supported by the Bavarian Academy of Sciences, Stromer conducted a second expedition to Egypt in the winter of 1903/1904 and finally a third in the winter of 1910. This third expedition was to become the most important of his career. He arrived in Alexandria in early November 1910, first to search for early mammals and then to explore the Bahariya area in the White

Desert, which consisted of Cretaceous sedimentary rocks. However, due to global political tensions on the eve of World War I,
it was difficult for a German explorer to get permission for such
an expedition from English, French, and Egyptian authorities.
However, because of his aristocratic status, financial resources,
and social influence, Stromer eventually succeeded in obtain-
ing the necessary permits. He met in November 1910 with the
British geologist John Ball, who had just created a geographic
map of Egypt and was in the process of publishing a geological
map of the country. Ball gave Stromer both maps, which were
invaluable to the paleontologist. After all, the area around Bahariya was hardly known at that time, and an expedition into
the western desert meant a dangerous adventure. For example, it
was always necessary to find new grazing grounds for the cam-
els, since the local workers had not bought fodder for the animals in advance, despite an agreement.

   After more than a week of hiking, Stromer's team finally arrived at the Bahariya Oasis. But because of the inaccuracy of the
geological map, he could not correctly assess the nature of the
rocks around Bahariya, and Stromer mistakenly believed that
the oasis dated to the Eocene, not the Cretaceous. In mid-
January, weather conditions improved, and he began to explore
the Gebel el Dist area. On the very first day, Stromer found a
fossil shark vertebra, fish teeth, and some petrified wood, and
four days later, he discovered three additional large bones that
he wanted to excavate and photograph. The first bone was a
femur, badly weathered and incomplete, but 43.3 inches long
and 5.9 inches thick. A second and better-preserved femur lay
next to it; it was 37.4 inches long and as thick as the other. The
third bone was stuck deep in the ground and could not be recovered at first. After that, Stromer discovered the pelvic bone
of a dinosaur, several vertebrae, and a gigantic claw. In all, he
found the remains of four very large, previously unknown dinosaurs in the Bahariya Basin. Three of them were carnivores:

Bahariasaurus, Carcharodontosaurus, and the famous *Spinosaurus aegyptiacus*. The fourth animal was a sauropod, which he later named *Aegyptosaurus*.

In addition to dinosaurs, Ernst Stromer and Austrian fossil hunter Richard Markgraf also excavated snakes, turtles, marine reptiles, and crocodiles there. However, since the excavation team had not brought any plaster, reinforcement fabric, or linen, Stromer had to improvise. Out of necessity, he cut up his mosquito net and soaked it in a mixture of flour and water to encase the two larger bones with it. The next day, he moved the expedition to Gebel Umm Hammad, where he also found several dinosaur, fish, and shark remains, and set out again two days later.

In February 1911 Stromer returned to Germany and over the next few years studied his finds. In 1912 he described *Spinosaurus aegyptiacus* and in 1931 another large carnivore, *Carcharodontosaurus saharicus*, which means "the shark lizard from the Sahara." Stromer chose this name because the animal's teeth had serrated cutting edges, reminding him of those of a great white shark, whose scientific name is *Carcharodon carcharias*. In 1932 followed the description of *Aegyptosaurus* ("the Egyptian lizard") and finally in 1934 that of *Bahariasaurus* ("the lizard from Bahariya").

Until 1914, excavations were still taking place around the Bahariya Oasis; then World War I broke out, and all the fossils that Stromer had found were held in Egypt. It was not until long after the end of the war, in 1922, that these finds reached Germany. They were badly damaged because they had been inspected with little care by the colonial authorities and stored improperly. The fossils probably would have been held in Cairo even longer if a Swiss paleontologist, a friend and former student of Stromer's, had not taken care of the paperwork and transported the fossils to Germany. Because of Switzerland's neutrality, this friend was able to talk to the British and French and get the pieces out of the country. Fortunately, *Spinosaurus*, Stromer's most important discovery, was not affected, as he had already taken it with him on his return trip from Egypt. The name *Spinosaurus* means "the spine lizard" and refers to the small, winglike projections of bone that point outward from each vertebra along the spine, also known as spinous processes. This animal was one of the largest predatory dinosaurs ever discovered, and we assume that the unusually long spines on its back likely had skin connecting them, forming a sail-like structure.

After the National Socialists rose to power in the 1930s, Stromer refused to join the NSDAP. He stood up for a Jewish colleague, incurring the displeasure of the party. When World War II began, Stromer's sons were drafted, and two of them fell on the battlefield. The third became a Russian prisoner of war and did not return home until five years after the war ended. Stromer tried in vain to convince the young and party-loyal curator of the Munich Museum of Natural History to remove the dinosaurs he had discovered in Egypt from the collection and bring them to safety. But the man refused because Stromer had publicly refused to swear allegiance to the Führer and he assumed out of arrogance that there would be no air raid on Munich. However, when the Royal Air Force bombed the museum on April 24, 1944, Stromer's collections were burned. The skeletons of *Spinosaurus*, *Carcharodontosaurus*, and *Aegyptosaurus* were completely destroyed. Ernst Stromer, whose life's work had been destroyed by the war, died a broken man in 1952. His work received little attention for several decades. Only when a new paleontological expedition to Bahariya was undertaken in 2000, and the remains of more dinosaurs were recovered there, did his research come back into the limelight. Among the new finds was the sauropod *Paralititan stromeri*, which bears Stromer's name in its honor. Whether the interest in his work also contributed to the appearance of *Spinosaurus* in the third Jurassic Park film, I cannot say. But I am sure that the paleontological advisors of the movie read the scientific description of the *Paralititan* and thought it would be cool to have the *Tyrannosaurus* and the *Spinosaurus* fight each other.

The question of how big and how heavy *Spinosaurus* actually was has long been the subject of great disagreement among experts. So far, no complete skeleton has been found, and the nature of its anatomy is hotly disputed. Most recent estimates, however, assume a length of about fifty feet, and there are hardly any deviations from this anymore. But as far as his weight is concerned, there is still great uncertainty. Depending on which calculations one follows, the animal is supposed to have weighed between four and up to twenty tons. More recent estimates are between six and seven and a half tons.

I am asked surprisingly often who would win in a real fight, the *Tyrannosaurus* or the *Spinosaurus*. I assume that this question is related to a scene from the third Jurassic Park film from 2001, in which a *Spinosaurus* effortlessly kills a *Tyrannosaurus*. Despite the fact that a time span of about 30 million years separates the two animals geologically, and that they would never have met geographically (*Tyrannosaurus* lived in North America and *Spinosaurus* in the Mediterranean), there is probably no doubt about the outcome of such a fight. *Spinosaurus* with its fifty feet was certainly somewhat longer than the approximately forty-foot-long *Tyrannosaurus*. However, the length gives only limited information about the mass of an animal. The extinct giant snake *Titanoboa* was probably much longer than most tyrannosaurs, but I don't think that a constrictor snake would have stood a chance against the king of the dinosaurs with his enormous teeth. It is similar with *Spinosaurus*. Its teeth—unlike those of most other carnivorous dinosaurs—were round in cross-section and were used more for snatching and holding prey than for cutting meat, biting tendons, or breaking and crushing bones and carapaces. The differences are striking. *Tyrannosaurus* teeth were longer, rooted deeper in the jaw, sturdier, and had cutting edges to bite huge chunks of flesh out of their victims. But the most obvious difference is the power with which a *Tyrannosaurus* could bite. Calculations have shown that it had the strongest bite force of any land creature ever. The bite force of a *Tyrannosaurus* was about three times as powerful as that of a *Spinosaurus*. And while *Spinosaurus*'s prey consisted of smaller or medium-sized animals it picked up while wading through lakes and rivers, *Tyrannosaurus* was capable of hunting and killing large herbivores, some of which were armored. Still, the two would probably have preferred to stay out of each other's way because, of course, there would still have been a risk of injury to *Tyrannosaurus* in a fight. However, if I had to make a bet on the outcome of such a fight, I would always bet on *Tyrannosaurus*.

Spinosaurus: On Land, on the Shore, or in the Water?

Spinosaurus is one of the most popular and well-known dinosaurs, and ever since new remains of it were discovered in the Kem Kem Basin in Morocco, researchers around the world have been taking a closer look at it. And because it's such an unusual dinosaur, distinctly different from other carnivores, discussions about its appearance, lifestyle, diet, size, and weight continue to inspire strange hypotheses. But I've been upset with the public perception of *Spinosaurus* and its portrayal in the media lately.

It all started with an article by Nizar Ibrahim and his colleagues in 2014 that described *Spinosaurus* as semiaquatic. There is a lot to be said for *Spinosaurus* living near water. It certainly stayed on the shores of streams and lakes, and its diet consisted to a not insignificant extent of fish. This is supported by the shape of its teeth, which resemble those of crocodiles and marine reptiles, and also by the shape of its skull, which is more like that of a crocodile than those of other carnivorous dinosaurs. In the article written by Ibrahim and other scientists, the team of authors attempted to prove an aquatic lifestyle by highlighting the fleshy nostrils located far back on the head, and the unusually long neck and torso for a carnivore, which shift the animal's center of mass well forward of its pelvis. In addition, the authors point out the short hind legs, which on land would have led to a disadvantage in hunting. Probably their most important argument is that the leg bones did not have an open medullary cavity and instead had a massive bony cortex on the shaft. Such bones are indeed indicative of a secondary adaptation to life in the sea, as we observe today in penguins, for example, which have very massive bones compared to other birds. If you look at the ribs in manatees and dugongs, you will notice that they are additionally broadened and thick and also lack a medullary cavity.

> **Secondary marine animals show such extra-wide ribs or thick limb bones without medullary cavity, one speaks of pachyostotic bones or pachyostosis. In my example, it is a secondary adaptation of manatees and dugongs to a life in water, acquired through evolution. In order to compensate for the static buoyancy of the body and to be able to remain effortlessly underwater, the average density of the animal increases, while the thickness of the bones increases, and the circumference of the medullary cavity decreases at the same time. The bones thus perform a function similar to a diver's belt.**

The team of authors succeeded in making a big splash on the topic; their article appeared in the prestigious journal *Science*, while *National Geographic* simultaneously published a long popular science article about their research. An Italian company made a life-size model of a *Spinosaurus* that was featured on the cover of *National Geographic*, and a Canadian company built a museum-quality skeletal model. Ibrahim gave a perfectly staged talk at the largest meeting of vertebrate paleontologists, and a movie was even made about how he serendipitously ran into a fossil dealer in Morocco who led him to a secret site in the Kem Kem Desert where new remains of *Spinosaurus* were found. The story was a huge media spectacle, and I seemed to be the only one who didn't take the bait. This was back when I was still working at the dinosaur park, where they were also in the process of making a new model of *Spinosaurus*. I remember well the discussions I had with the park's general manager, the sculptor, the scientific director, and a former fellow student from Bonn who was now also working in the park. They all found Nizar Ibrahim's article interesting and wanted to build the model according to his specifications. My criticisms fell on deaf ears.

A heated exchange with the managing director, in which he pointed out to me that four votes were in favor of these changes and only one vote (mine) was against them, was the last straw. I replied undiplomatically that science is not a democracy and only the facts count. After that, it was clear that I would not be working in the park much longer.

What bothered me was the fact that Ibrahim's reconstruction was based on a composite of several individuals, whose proportions I thought he had wrongly scaled and misinterpreted. The animal looked kind of jumbled together even at first glance.

The second argument that made no sense to me was the reference to the pachyostotic bones. I had already seen, during excursions to the Mainz Basin, fossils of sea cows that had swam around with their thick, broad ribs in the shallow sea there during the Oligocene, about 30 million years ago; moreover, I had observed pachyostosis on a primitive mammal, the enigmatic *Desmostylus* in the collection of the Tsukuba University in Japan, and therefore I knew that the bones of *Spinosaurus* were quite different in nature.

Apart from that, it was absolutely incomprehensible to me why the leg bones in an animal should be thicker, but at the same time shorter, to give it additional weight. On the contrary, this animal would have to become rather lighter by such a development, because the mass gain by pachyostotic bones would not have outweighed the mass loss of smaller muscles on the smaller legs.

Also, I did not see any significant widening of the ribs, which would have been much more important for static lift. And why, of all things, would an animal with such a massive dorsal sail live in the water? A crocodile ambushing unsuspecting wildebeest or zebra at a water trough might be completely submerged until only its nostrils peek out of the water. In the same situation, a spinosaur's dorsal sail would have protruded out of the water by up to six feet. This would certainly not have been a good strategy for an ambush predator, whose hunting success heavily depended on the surprise effect. If *Spinosaurus* had then also decided to swim or submerge so far that its sail would also have been underwater, the water resistance would have made any change of direction impossible, and the animal would have drifted helplessly as soon as the sail accidentally got crosswise to the direction of flow.

Ibrahim and his colleagues only marginally addressed the sail and its hydrodynamic properties. Instead, they argued that this body part was probably a display feature males used to make an impression on females. I don't doubt that, because on land such an imposing structure would certainly have caused a stir, but in the water, it would only have been a hindrance.

I don't blame the media here—the whole story was sensational and exciting, because *Spinosaurus* would have been the first non-avian dinosaur adapted to a life in water. Dinosaurs with such a high degree of adaptation to water are only found among today's birds. Penguins are the only birds that show quite extraordinary adaptations to life in the sea.

To test Ibrahim's hypotheses, Donald Henderson created three-dimensional digital models of *Spinosaurus* and other animals in 2018 that also accounted for regional density variations in

the animals' lungs and air sacs. In doing so, Henderson noted that
the model of *Spinosaurus* tended to tip on its side in the water.
His models of *Baryonyx*, *Tyrannosaurus*, *Allosaurus*, *Struthiomimus*,
and *Coelophysis* showed similar buoyancy, suggesting that *Spinosaurus*'s hydrodynamic properties were far from exceptional.
The software also provided evidence that the center of mass of
Spinosaurus was close to the hip. This would allow *Spinosaurus*
to move forward well on land. Its pneumatized skeleton and air
sacs ensured that it was unsinkable. A semiaquatic lifestyle of
pursuit hunting in the water is thus ruled out. Henderson concluded that *Spinosaurus* may have been more specialized for a life
on the shoreline, wading only occasionally in shallow water. At
the same time, it may have been a competent hunter on land,
spending a significant portion of its life away from the water.

The article about the static buoyancy of this dinosaur did not
receive nearly as much attention as the media spectacle about
the swimming *Spinosaurus*. After Henderson's article appeared, I
pointed out his new findings in an online forum and also mentioned the unfavorable hydrodynamic properties of the large
dorsal sail. I must admit that this was somewhat naive, as my
contribution was immediately criticized, and I was personally in-
sulted and responded to with hostility. I don't normally converse
in these forums, so I was unfamiliar with the strange dynamics
of such groups, where anyone can just post anything without
backing up their claims with appropriate scientific evidence. I
was asked who I was that I could dare to doubt the "fact" of an
aquatic lifestyle of *Spinosaurus*. When I looked at the profile of
my vociferous critic, to my knowledge this person had no academic background, nor did he practice any scientific profession.

Though this topic is heavily debated and there are many op-
posing theories, I don't understand at all why everyone seems
to want to make *Spinosaurus* into a creature that it is not. With
its massive size, huge dorsal sail, and crocodile-like snout, *Spinosaurus* is one of the most fascinating dinosaurs, whether it was

aquatic or not, and its popularity today almost rivals that of *Ty-rannosaurus*. Unfortunately, the movie *Jurassic Park III* and the 2014 publication previously mentioned have contributed to the public's misconception of *Spinosaurus*. It was such an extraordinary animal that there is no need to attribute additional superpowers to it. In 2020, Nizar Ibrahim published a new study to debunk Henderson's arguments and provide more evidence of strong aquatic adaptation. In principle, I think highly of maintaining discourse on a controversial topic when new evidence emerges. However, sometimes one has to wonder if new work actually adds to our scientific understanding or has been published for questionable reasons. While Ibrahim and his team acknowledged that Henderson had presented compelling anatomical, biomechanical, and taphonomic reasons to question a semiaquatic lifestyle, they also presented evidence for a putative swimming tail. Their article appeared this time in the prestigious journal *Nature*. They argued that *Spinosaurus* possessed extremely high dorsal and ventral spinous processes on its caudal vertebrae, forming a large and flexible finlike organ with which it could supposedly perform wide-ranging swimming movements. To demonstrate these swimming characteristics, they built several underwater models of oscillating tails, measured the physical forces on them, and compared different tail shapes. They wanted to show that the tail of *Spinosaurus* produced greater thrust in the water than that of other supposedly land-dwelling dinosaurs, and that *Spinosaurus* was therefore more comparable to aquatic vertebrates, which had laterally flattened and vertically extended swimming tails to produce propulsion while swimming. Unfortunately, the story was again inconclusive from my perspective. Namely, the team claimed that all *Spinosaurus* relatives had the characteristic crocodile-like teeth and elongated snout, suggesting that they were all piscivores, yet the tails of the remaining spinosaurids showed only minor aquatic adaptations.

What bothered me about this experimental setup was that the two-dimensional robotic models of the swimming tails could only inaccurately simulate the movements of a three-dimensional body in the water. This was also noticed by David Hone, a colleague from London, and Thomas Holtz, one of the world's leading experts on predatory dinosaurs, who took a closer look at Ibrahim's study. They included new finds of *Spinosaurus* in their considerations. The two paleontologists evaluated arguments about the animal's functional morphology, ecology, and its hunting and feeding habits in the context of these new finds. They were convinced that the degree of Ibrahim's aquatic adaptations was overestimated. The interpretation of the anatomy of the tail as a caudal fin was not tenable in their view, and they expressed considerable doubt that *Spinosaurus* was able to swim efficiently and rapidly with this tail and to pursue prey underwater. They saw no evidence for a highly specialized adaptation as an aquatic predator and instead proposed a model that *Spinosaurus* fished predominantly along the coast or in shallow waters but could also search inland for prey. Even evidence that *Spinosaurus* was a good swimmer would not be counterevidence for a wading lifestyle and the fact that this dinosaur may have sought its food along the coast. Conversely, the instability in the water, the high water resistance, the position of the eyes and the nostrils, and the low swimming performance argued against an efficient hunter underwater. Another aspect that had been noticed in previous studies was the underside of its neck, which had rugose muscle attachments, indicating strong musculature that may have been useful for rapid, downward pecking movements of the head. The animal may have fished in the water with its snout, as storks do today. It may have remained underwater with its mouth slightly open until a fish swam by, and then snapped at it, or lunged at the prey with a rapid downward movement. The ability for such rapid downward movements would not have benefited a pursuit predator underwater because there, fish can escape in all directions. In this context, the position of its nostrils far back on the skull is advantageous, since *Spinosaurus* could still breathe easily even with its mouth underwater, as storks can. David Hone and Thomas Holtz therefore drew a picture of a generalist that searched for prey along the shore, in and out of the water, and also did not disdain meat of animals that had come to their death at watering holes or during floods. The denser leg bones would have aided its wading because they would have minimized unwanted buoyancy in the water. By being able to hunt in the water and on land, there was less competition for these animals from crocodiles and land-only predators, and they could also move more easily from one lake to the next.

There is another piece of evidence that supports this hypothesis. Examinations of an animal's tooth enamel can show what it feeds on predominantly. Certain isotopes reveal whether it ate more land animals or fish. In the case of *Spinosaurus*, this signal is ambiguous because it probably made its prey both on land and in the water.

In late 2022, Professor Paul Sereno of Chicago joined the discussion about *Spinosaurus*. Ibrahim was Sereno's student and Sereno had been one of the coauthors of his 2014 article. Therefore, I was surprised when Sereno criticized Ibrahim's second *Spinosaurus* article from *Nature* and came to a completely different conclusion. Paul Sereno and his colleagues used computed tomography to scan all available parts from different specimens of the animal and created a three-dimensional model to which they added muscles, air sacs, and lungs to model the weight distribution as realistically as possible. They showed that *Spinosaurus* was only about forty-five feet long and, while on land, walked bipedally rather than on all fours, while its body was laterally

unstable in deep water. It was a slow surface swimmer with weak
tail muscles that could not move forward in water faster than
2.2 miles per hour. Said paleontologists also demonstrated that
Spinosaurus's static buoyancy was so great that it certainly could
not dive. This dinosaur could have waded to a water depth of
about 8.5 feet but would then have floated on the surface of the
water, unable to submerge. In fact, while its hind legs had slightly
thicker bone walls, its dorsal vertebrae had air-filled cavities in
front of the pelvis, and the bones of its forearms had a pronounced
medullary cavity. Sereno and his colleagues calculated that because of the strong pneumatization of the vertebrae, *Spinosaurus*,
despite its considerable weight of about 7.4 metric tons, had a low
density of only about fifty-three pounds per cubic foot, which
would have been far too little to dive on its own.

> **Salt water has a density of about sixty-four pounds
> per cubic foot.**

Sereno's team pointed out that reptiles living today with similar dorsal sails or high, laterally flattened tails, such as the common basilisk (*Basiliscus basiliscus*), use them for display rather than
actual aquatic locomotion. Vertebrates secondarily adapted to
aquatic life, on the other hand, almost invariably have a fleshy
fluke or caudal fin rather than a flattened swimming tail, and
also have greatly reduced limb length. *Spinosaurus*, with its powerful legs, on the other hand, could reach far inland without
using its long-clawed hands for weight support. Although its
physique suggests a semiaquatic lifestyle, the research team ruled
out a purely aquatic lifestyle because at no time in the fossil record were there secondary aquatic vertebrates that lived in fresh
water and were longer than seven feet. Such animals all lived in
the open ocean, far from the coast. A semiaquatic lifestyle, on
the other hand, can occur at any body size.

Fish and No Chips:
The Fish Eaters of the Isle of Wight

Ernst Stromer discovered the first remains of *Spinosaurus* in North Africa, but only recently have we learned how diverse the spinosaur group is, through finds on an island in the English Channel: the Isle of Wight. In 2010, a paleontological field trip from the University of Bonn under Professor Sander took me there for the first time. I immediately fell in love with the island because of its beauty and stunning dinosaur fossils. You can only reach it by ferry from Portsmouth or Lymington, which dock in the north of the island. However, dinosaurs and their footprints are found on the Isle of Wight only in the southwest and southeast. The rocks in which their fossils are found belong to the so-called Wealden Group; they date from the Lower Cretaceous and are about 125 to 140 million years old.

PHOTO BY THE AUTHOR

At the south coast of the Isle of Wight, United Kingdom, natural casts of dinosaurian footprints are exposed.

The south coast of the Isle of Wight is very steep and experiences strong erosion. There, fossilized footprints of dinosaurs crop out, which are gradually washed into the sea. If they are not salvaged, they will eventually be swallowed up by the waves. Such footprints were formed when a dinosaur walked on the soft clay or sandy ground and sank a bit into the subsoil, leaving imprints. These impression molds were then infilled with sand or clay shortly after their formation and were thus protected. This allowed both the soil and the infillings, called casts, to slowly turn into stone. Sometimes the rock that forms these casts is harder or more resistant to weathering than the surrounding matrix, and when the matrix weathers away, millions of years later, only the casts remain. Sometimes these casts are not quite clearly identifiable as such, but if you have an eye for it you can easily spot them. On the Isle of Wight, footprints of dinosaurs are found far more frequently than their bones. A single imprint in sedimentary rock is simply called a footprint, while several footprints in succession are a track, and the layer on which such a track is found is a track site.

The most common tracks found on the Isle of Wight are those of iguanodontids. *Iguanodon* was one of the largest ornithischian dinosaurs of the Lower Cretaceous. After *Megalosaurus, Iguanodon* was the second dinosaur to be scientifically described, in 1825. Particularly striking are its bony thumbs, which were initially misinterpreted as a nasal horn. The first reconstructions show *Iguanodon* with a quadrupedal, or four-legged, mode of locomotion. Later, it was depicted as bipedal for nearly a hundred years. But more recently it is assumed that it actually walked mainly on all fours and only moved bipedally when it fled from danger. Tracks of *Iguanodon* confirm this assumption. Since its hoof-like forefeet were much smaller, it is easy to distinguish impressions of their hands from their feet. Another herbivorous ornithischian dinosaur found on the island is *Hypsilophodon*. It was significantly smaller and actually a nimble biped.

Iguanodon and *Hypsilophodon* possessed three toes, the outer two of which were splayed at a ninety-degree angle. Predatory dinosaurs have very similar tracks to these two animals, but their toes are usually more slender, and the outer toes form an angle of less than ninety degrees. The impressions of predatory dinosaurs on a track site sometimes also retain the hollow shapes of the sharp claws, which facilitate assignment. In Germany, for example, such footprints can be found at the Dinosaur Park Münchehagen. The sediments in which they occur are as old as those on the Isle of Wight. The special thing about the dinosaurs on the English island, however, is not only the high number of finds, but above all the incredibly diverse dinosaurian fauna. Even long-necked dinosaurs such as *Cetiosaurus* and *Ornithopsis* were found there. A particularly complete skeleton of a long-necked dinosaur is the so-called Barnes High sauropod. In 1992, almost its entire skeleton, including the head, was found, which is very rare in sauropods. The animal still has no scientific name, but it was closely related to *Brachiosaurus* and is considered the most complete sauropod skeleton of England. There are also remains of armored dinosaurs on the Isle of Wight, such as *Polacanthus*, whose name means "many thorns." It is an early representative of the ankylosaurs that carried spines along their flanks and bony plates—called osteoderms—on their backs. These osteoderms were fused together over the pelvis to form some kind of armor. With all these finds, it is not surprising that the Isle of Wight is also called the Dinosaur Island of England. More than twenty-five different species of dinosaurs are already known from there and that doesn't even include the undescribed finds. When these species are also eventually described, the number will increase significantly. What makes the dinosaur fauna of this island so interesting for research is the especially high number and diversity of predatory dinosaurs.

PHOTO BY THE AUTHOR

The Needles form the western tip of the Isle of Wight and are the main tourist attraction of the island.

With a body length of up to twenty-two feet, *Neovenator* was the largest purely carnivorous dinosaur on the Isle of Wight. It could move quickly and had three deadly claws on each foot and razor-sharp teeth in its mouth. Unlike *Tyrannosaurus*, however, these teeth were very thin and better for cutting meat than for crushing or biting bone. Tyrannosaurs also play a role in the fauna of the island. In fact, the oldest representative of the group and thus the earliest ancestor of *Tyrannosaurus rex* was discovered on the Isle of Wight. Its name is *Eotyrannus*, which means "the tyrant of the dawn." He was about sixteen feet long and was described only in this century. Even bigger than *Neovenator* was *Baryonyx*. It was probably not a pure carnivore, but rather a fish-eater, as suggested by its teeth and its crocodile-like skull shape. *Baryonyx* was a close relative of *Spinosaurus* but did not have a dorsal sail. Its skeleton is the most complete of any spinosaur to date. In 2021, two new spinosaurids were described: *Riparovenator*—"the hunter of the riverbank"—and *Ceratosuchops*, "the horned crocodile face." However, not many remains of these two animals have been found. In 2022, another spinosaurid was reported. It too appears to be a new species, as its fossils were found in younger strata, the Vectis Formation on the southwest coast of the island, near the village of Compton Chine, which—strictly speaking—consists of only a farm and a large parking lot. The find contains no skull bones, which is why researchers have not yet assigned a scientific name to the animal. Cervical, pelvic, and caudal vertebrae are well preserved, as well as parts of the ilium, ribs, and fragments of arm or leg bones. These remains suggest an individual of enormous proportions, comparable to the size of the *Spinosaurus* found by Stromer. This would make the animal much larger than the other known spinosaurs from the island. And, assuming similar anatomy to *Spinosaurus*, this find could even be the largest carnivorous dinosaur ever found in Europe. A comparison with the bones of other spinosaurids shows that this yet unnamed animal is more closely related to *Spinosaurus* than to the other spinosaurids of the Isle of Wight, such as *Baryonyx*, *Ceratosuchops*, and *Riparovenator*. The find from the Vectis Formation also suggests that spinosaurids diversified and occupied new ecological niches since their origin. While the earlier representatives from the Wessex Formation predominantly inhabited the banks of extensive river systems, this animal apparently lived near a lagoon from which few other dinosaurs are known to date.

What fascinates me is that spinosaurids probably always occurred in ecosystems that had very high carnivore diversity. We see that with Ernst Stromer's dinosaurs from the Bahariya Oasis, which included *Spinosaurus*, *Carcharodontosaurus*, and *Bahariasau-*

rus. And the same is true on the Isle of Wight, where, besides the three spinosaurids *Baryonyx walkeri*, *Ceratosuchops inferodios*, and *Riparovenator milnerae*, there may have been a fourth, much larger genus from the same family, plus the tyrannosauroid *Eotyrannus lengi*, the large predator *Neovenator salerii*, and a number of other carnivores whose position in the dinosaur family tree is not entirely clear.

In Morocco we have a similar picture. There lived besides *Spinosaurus* also *Carcharodontosaurus* and huge crocodiles, which were so big that they probably ate even dinosaurs. And on the Isle of Wight, the fragmentary remains of the new spinosaurid suggest that it too may have been one of the largest carnivores in Europe.

We will learn later, in the chapter on *Tyrannosaurus*, that this was very unusual for dinosaurian faunas. And the fact that in these assemblages oft particularly large predators occurred, makes little sense in regard to the distribution of large predators within the trophic network.

CHAPTER 8

More than Just Bones

The Super-Lung

Through the various fossil discoveries, we have learned much about the powerful lungs of dinosaurs. We see evidence of unidirectional lungs in the highly pneumatized bones of the vertebrae, and we can infer their functioning from the anatomy of modern birds and crocodiles. Many of these conclusions are based on computer models and theoretical considerations. Yet for a long time there was no clear, direct evidence because the lung, like all other soft tissues, has a low preservation potential. But that changed when an animal called *Archaeorhynchus* was reported from China that was already more closely related to modern birds than to enantiornithines. It came from the Jiufotang Formation of the Lower Cretaceous, and its soft tissue and plumage were still very well preserved. In its thorax were the remains of two lungs that resembled those of living birds, an indication that even these early forms of birds were capable of flight 120 million years ago. Of all living, air-breathing vertebrates, birds have the most complex and efficient respiratory system, allowing them to maintain their energetically demanding form of locomotion even in oxygen-deficient environments. The lung microstructure of *Archaeorhynchus* also appears modern and is further evidence that many physiological changes in the respiratory system that characterize extant birds and helped them succeed were already present in dinosaurs that were not directly on the avian lineage. Instead of saying that dinosaurs possessed birdlike lungs, it would probably be more accurate to say that birds possess dinosaur lungs.

The Cloaca:
How Do Dinosaurs Make Wee-Wee?

Walking through the halls of the Senckenberg Museum in Frankfurt, it's easy to lose track of all the treasures the museum has to offer. In the large hall, skeletons of a long-necked dinosaur, a horned dinosaur, an iguanodontid, a club-tail dinosaur, and a tyrannosaur are on display. Behind it lies the famous dinosaur mummy. In this setting, this original fossil of a small herbivore hardly catches the eye, lying inconspicuously near the wall in a display case. It is the Frankfurt specimen of a *Psittacosaurus* from the Early Cretaceous Jehol deposits of Liaoning in China. This fossil has the best preservation of scale-covered skin of any ornithischian dinosaur described to date. One can even discern its color pattern and countershading, which allows us to reconstruct in detail the appearance of this animal and draw conclusions about its habitat. The countershading suggests that the animal lived in a shaded, forested environment. Because the skin and scales are so well preserved, this piece also allows us to see a cloaca on a dinosaur for the first time.

The Psittacosaurus *specimen of the Senckenberg Museum in Frankfurt, Germany, shows remarkable soft tissue preservation.*

A cloaca is the posterior orifice that serves as the only opening for the digestive, reproductive, and urinary tracts. It was originally present in all vertebrates but was replaced by separate excretory orifices in bony fish and most placental mammals. The cloaca is the posterior section of the rectum, to which the reproductive organs and the ureters are connected. Through it, sperm, ova, and excrement are discharged via the anus.

The fact that dinosaurs had a cloaca is basically no surprise, because it is also true for all archosaurs living today, even if it looks different in birds and crocodiles.

To better compare the small dinosaur with its living relatives, the specimen containing the fossil was scanned, which allowed retrodeformation of its virtual model. The preservation of the rectum of this *Psittacosaurus* is remarkably good and shows in detail the anatomy of its cloaca. Its opening is clearly visible and can be readily distinguished from surrounding tissues, as the shape of the scales and pigmentation differs from those of adjacent body regions. Although the cloaca gives no indication of the sex of the animal, the pigments appear to have been signal colors, possibly intended to create a visual stimulus. One might think of it as in baboons or mandrills, whose rumps are colored differently than the rest of their fur.

Do Female Dinosaurs Have Ovaries?

Hard-shelled eggs are a characteristic shared by all birds and crocodiles. Yet the reproductive systems of all living archosaurs are very different—as are their cloacae. Crocodiles have two ovaries, their ovarian follicles mature slowly, and they have large clutches and small eggs. Living birds, on the other hand, almost all have only one ovary and one oviduct, and their follicles develop rapidly. Birds are now the most diverse tetrapod class, and there is a wide range of clutch and egg sizes within their group, depending on the body size of the bird. Generally, however, the eggs of birds are larger, and their clutches are smaller than those of crocodilians.

Derived reproductive traits in birds are thought to have evolved gradually early in the dinosaur evolution, but the exact timing of these changes is difficult to trace because of the incompleteness of the fossil record and because of sparse soft tissue finds. However, we have learned much from fantastic finds from the Jehol Province in northeastern China. Recently, there have been reports of exceptionally well-preserved ovarian follicles in the prehistoric bird *Jeholornis* and in several enantiornithines from the Early Cretaceous strata of this region. Initially, it was doubted whether these were actually follicles or rather seeds that had been eaten by the animals.

To find out, fragments of these alleged follicles were extracted and subjected to several analyses. These structures were found to have histological and histochemical characteristics of smooth muscle fibers intertwined with collagen fibers, resembling the contractile structure of connective tissue in birds today. Fossilized blood vessels were also preserved. No plant fibers or evidence for other plant tissue was found, supporting the original interpretation as follicles in the left ovary. At the same time, the right ovary was absent, showing that it was apparently functionally lost early in avian evolution. The fossils from China were compared with the small carnivore *Compsognathus* from Germany, which, like *Archaeopteryx*, was found in the Solnhofen Limestone.

As it turned out, *Jeholornis* and *Compsognathus* both show features similar to those of present-day crocodiles because they still have a large number of small ovarian follicles. Enantiornithines also differ from modern birds in terms of ovarian follicles. They had fewer and larger follicles; however, these follicles were all similar in size. In today's chickens, the size of the follicles in the ovary varies, corresponding to different stages of development.

The Grace of the
Sleeping Club-Tail Dragon

Kai Jäger, one of my former fellow students from Bonn, dealt in his research with the evolution of early mammals. Kai can talk about his work in a very lively and humorous way, thus making it more accessible to a broad audience. That's why he also took part in Science Slam, a competition in which researchers have to present their results as entertainingly as possible in a very short period of time. The focus is on communicating scientific content in a way that is broadly accessible. The participants' ten-minute presentations are voted on by the audience through cheering with the winner being the one who receives the most applause. Kai did so well that he was invited to the German Science Slam Championship in Berlin, which he actually won in 2014. Since then, he has been regularly invited to speak by radio and television broadcasters when an expert in paleontology is needed. Hence, it was not surprising when in 2017 the Deutschlandfunk radio station called him when a spectacular find of a

club-tail dinosaur was reported from Canada. However, he forwarded this request to me, since he is not a real dinosaur expert,
but as a paleontologist is mainly interested in the evolution of
mammals. This radio program was about an ankylosaur that had
been discovered in the oil sands of Canada, and since I had already read quite a bit about it, I was familiar with the topic. The
animal, named *Borealopelta markmitchelli*, was found in a mine in
northeastern Alberta. There, workers came across its skeleton in
a debris heap, and employees of the mining company immediately informed the Royal Tyrrell Museum in Drumheller, one
of the best-known and most famous paleontological museums
in the world. It has one of the largest and most speciose collections of dinosaurs, with an emphasis on fossil finds from the Canadian Badlands. A museum team traveled directly to this mine,
recovered the fossil with help from the workers, and shipped it
back to Drumheller. The rock strata in which the animal was
found is composed of Albian marine sediments and belongs to
the Clearwater Formation, where several plesiosaurs and ichthyosaurs were also deposited. However, no dinosaur had ever
been discovered in it until that time, which is not surprising because the formation contains sediments of a coastal facies and a
nearshore marine environment. Normally, land creatures would
not be present in these sediments. Most likely, said *Borealopelta*
had fallen into a river and had been washed away by it. Whether
this happened before or after the death of the animal is not con-
clusively clear. In any case, after its death, the carcass, floating
on its back, entered the open water and then sank to the bottom
of the sea. There, the animal was quickly buried by sediment.
Although remains of seafloor benthos such as worms, snails, and
starfish can be found in these deposits, there was no evidence of
feeding marks on the dinosaur by such scavengers. If, despite the
seafloor being oxygen-rich and populated by burrowing animals, the carcass does not show any feeding marks, it is an unmistakable sign of a very rapid burial. When the fossil was found,
it was completely encased in a dense and hard—but at the same
time brittle—iron carbonate concretion, such as the one we have

already encountered with the ichthyosaurs from Nevada. Cracks on the surface of the concretion showed that it must have fractured as the carcass collapsed in its calcareous grave, as the organic material increasingly decomposed, liquefied, and produced decomposition gases. When the loading pressure finally became too great, the gases escaped, and the body fluids were released. The resulting cavity in the dinosaur's carcass was then filled with sediment. However, because the animal was lying underwater with its belly side up, its carapace remained intact. As a result, not only have its scales and osteoderms been almost completely preserved, but so have their arrangement in their natural compound, and—much more unusual—many of the dermal bone plates are still covered by a horny sheath. Because the empty body cavity was so quickly filled with sand, the fossil is barely crushed, and we now have an excellent three-dimensional preservation. The Canadian research team was even able to identify the pattern of its scales and systematically searched for melanin pigments which were preserved on the carapace.

> **Melanin are widespread dark brown to black or sometimes yellowish or reddish pigments that cause coloration of the skin, hair, feathers, or eyes in animals.**

To reveal all of this, the fossil had to be freed from the concretion in the laboratory first, which took the preparator Mark Mitchell a full five years and around seven thousand hours of work. His extraordinary achievement was recognized with the naming of the animal, whose species name ended up being *markmitchelli*. The genus name *Borealopelta* is a combination of Latin and Greek and means "the shield from the north," alluding to its northern location and its carapace. Mark Mitchell's effort was clearly worth it, because the fossil is very beautiful. For its display, they turned it upside down so that the carapace was again facing up. If you look at it from the front, you might think *Borealopelta* is still alive and just sleeping, which has earned it the nickname "Sleeping Beauty." Although the animal may not be as pretty as a king's daughter, at least it looks like a sleeping dragon, with its entire front part preserved. The head, neck, most of the torso, and its limbs were protected by osteoderms during its lifetime. On its flanks, the animal also bore long spines. One might think it was a walking tank on four legs. But the color pigments of its carapace reveal that even adult *Borealopelta* must have had predators. The pigments are distributed over its carapace in a countershading pattern and were meant to fool predators.

> **When the coloration of an animal is darker on the upper side and lighter on the lower side of the body, it is called countershading. Because this phenomenon was first observed and described by the painter Abbott Thayer, it is also called the Thayer's principle. Countershading is a form of camouflage because it makes it more difficult for predators to see their prey.**

Countershading was already observable in ichthyosaurs, and we will encounter this survival strategy again in other dinosaurs in a later chapter. When I was asked about this in the radio interview on Deutschlandfunk, I expressed doubts about the interpretation of the carapace coloration of *Borealopelta*. I was of the opinion that *Borealopelta* did not need camouflage because it was simply too large and too well armored. Even large tortoises, where small members of the group still have camouflage colors, lose their coloration without predation pressure. With an estimated length of eighteen feet and a body mass of at least twenty-nine hundred pounds, *Borealopelta* was much larger than any modern land animal that exhibits countershading. In today's mammalian faunas, predators pose no risk to herbivores weighing more than a ton. These huge animals do not need to hide and therefore have no camouflage. If they also have defensive weapons such as horns or tusks, predation pressure decreases even further. So why would *Borealopelta* have needed camouflage? I found it more plausible that its reddish coloration represented a form of aposematism (advertising signals in the form of conspicuous coloration) that some venomous animals display. I could also imagine that the strong coloration might have played a role in mating and courtship. However, I had not considered that the predator-prey situation was completely different in the Cretaceous. While there are no remains of carnivores in the same strata, we do know of footprints of particularly large predators in the Clearwater Formation from descendants of allosaurs. Weighing several tons, these theropods were strong enough to kill even an adult club-tail dinosaur; they were more than ten times heavier than the largest predators of the present day (such as tigers, lions, and polar bears). We'll learn more about the difference between dinosaur- and mammal-dominated ecosystems when we get to tyrannosaurs. But the hunters of *Borealopelta* were not tyrannosaurids. In fact, they didn't appear until about 50 million years later.

The fact that camouflage was especially important in dinosaurs, crocodiles, and birds is apparently also related to the exceptionally good eyesight of modern birds and crocodiles. We assume that these groups inherited the ability from their common ancestors, and therefore assume that the large theropods also had very good vision and that these carnivores were very visual predators. This is in contrast to modern ecosystems, where large mammals at the top of the food chain can only perceive a limited spectrum of colors, even when compared to humans. The enormous size and strong visual dependence of Cretaceous apex predators may have led to an evolutionary arms race that resulted in a combination of armor and camouflage even in the largest herbivorous dinosaurs. Thus, club-tail dinosaurs also had to hide and camouflage. The presence of countershading in a large, heavily armored herbivorous dinosaur thus offers us a unique glimpse into a Cretaceous predator-prey dynamic that is not comparable to processes in nature today.

THE UPPER CRETACEOUS

(100.5 to 66 million years
before present)

Edmontosaurus

Tyrannosaurus rex

Triceratops

Denversaurus

Gilmoremys

Potamornis

CHAPTER 9

Wyoming—The Hell Creek

Triceratops: *A Tank on Four Legs*

One of the most important paleontologists of our time, Professor Roger Benson, once said that there was one correct answer to the question of which dinosaur was the best: *Triceratops*!

That was just a joke, of course, but I was pleased anyway, because *Triceratops* has always been my favorite dinosaur. Even in my childhood, I enthusiastically gazed at the skulls of *Triceratops* in Frankfurt's Senckenberg Museum. My poor brother unfortunately had to be there with me, too, although he was not at all a dinosaur fan. Over the years I've seen many skeletons, but I always like to return to Frankfurt to take a selfie with *Triceratops*. The animal is on display in Frankfurt, although it is actually from Wyoming. *Triceratops*, along with *Tyrannosaurus rex*, was one of the last dinosaurs outside the avian lineage. Remains of *Triceratops* and *Tyrannosaurus* are found in the same strata. They come from the Lance Formation and the Hell Creek Formation, which were deposited during the last 2 million years of the Cretaceous period.

The two skulls of Triceratops, *discovered by Charles Sternberg, now on display at the Senckenberg Museum in Frankfurt, Germany.*

Interest in *Triceratops* skeletons has been high in recent years, with several auctions of these fossils causing a stir. In October 2021, one particularly massive specimen sold for 6.65 million euros (about 7.7 million US dollars at the time) to a private collector at an auction in Paris. At the time, people worried that the animal would no longer be accessible to the public and to researchers in the future. Fortunately, however, the fossil found its way to the museum of the D'Annunzio University of Chieti–Pescara in Italy. The giant skeleton was discovered in South Dakota in 2014 in the Hell Creek Formation. It is about 23.5 feet long and has a hip height of almost nine feet. Because of its size, it is also nicknamed Big John. The skull alone accounts for more than one third of the animal's total length. It is the largest nearly complete *Triceratops* skeleton in the world; 60 percent of the body and over 75 percent of the skull have been preserved. However, we know of partial skeletons and isolated skulls that suggest there must have been much larger specimens. The word *triceratops* means "three-horned face." This dinosaur had two long horns over its eyes and one on its nose. It belongs to the group Certatopsia, or horned dinosaurs. Almost everyone knows the name *Triceratops*, but surprisingly few people know that this dinosaur, just like *Tyrannosaurus rex* and every other scientifically described animal, also has a species name. There are two valid species names of *Triceratops*: *Triceratops horridus* and *Triceratops prorsus*. *Horridus* (Latin for "rough" or "wrinkled") refers to the uneven surface of its bones and *prorsus* (Latin for "forward") refers to the forward-pointing nasal horn. A total of seventeen different species of *Triceratops* have been described, with all but the two mentioned now invalid. When horns of a *Triceratops* were first discovered, they were thought to be the horns of an extinct species of bison and the find was named *Bison alticornis*, meaning "bison with high horns." If you count this name, the *Triceratops* has eighteen species names.

The two valid names of *Triceratops*:
T. prorsus, T. horridus
Invalid names used as synonyms for the valid species:
T. albertensis, T. alticornis (originally *Bison alticornis*), *T. brevicornus, T. calicornis, T. elatus, T. eurycephalus, T. flabellatus, T. galeus, T. hatcheri, T. ingens, T. maximus, T. mortuarius, T. obtusus, T. serratus, T. sulcatus, T. sylvestris.*

Triceratops was first described by Othniel Charles Marsh in 1889. The horns, which he thought were those of a bison, came from the Denver region of Colorado. The first whole skeleton came from the Lance Formation in Wyoming. *Triceratops* bones were also found in the Hell Creek Formation. This dinosaur was hunted by the most dangerous predator of all time, *Tyrannosaurus rex*. But *Triceratops* was also an extremely strong and dangerous animal and by no means defenseless. *Tyrannosaurus* could only attack by ambushing *Triceratops*. It is quite possible that some tyrannosaurs did not survive their attack on a *Triceratops* because they themselves were seriously injured in the process. The bony, three-foot-long horns above the eyes of *Triceratops* were dangerous weapons. Like those of cows and buffalo, they were covered with a horn sheath made of keratin. The animal had a thick, massive frill that protected its delicate neck, shoulders, and nape. *Triceratops*'s head was connected to its spine by a ball-and-socket joint, so that it could turn in all directions at lightning speed and an attacker always had to stare at its pointed horns first. If *Triceratops* was not caught by surprise, it was very difficult to kill it. The first real evidence of a duel between a *Tyrannosaurus* and a *Triceratops* was provided by a spectacular fossil discovered in Montana in 2006, which is now on display in Raleigh, North Carolina. The fossil contains a not-yet-fully-grown *Tyrannosaurus* and a *Triceratops* that are preserved, joined in mortal combat. The young *T. rex* obviously overestimated itself in the selection of its prey, and in the end both adversaries came to death. *Triceratops* was rather quarrelsome and frequently engaged in territorial and mating fights, proven by its injuries on the frill of Big John. It bears a puncture wound that, judging by its shape and size, was inflicted by the horn of a conspecific. Although the wound had already partially healed on the edges, it must have become infected while the animal was still alive. Perhaps it was so deep that Big John later died from the infection. However, the neck shield of *Triceratops* served not only as a defense, but also as a means of attracting a mate. It is quite possible that these shields had conspicuous display colors. Some scientists initially even thought that horned dinosaurs developed their frills to distinguish between different horned dinosaur species and to separate their own species from others, but this has since been ruled out. *Triceratops* had an additional bone in the lower jaw that is found only in ornithischian dinosaurs such as duckbill dinosaurs and horned or armored dinosaurs. This bone lies anterior to the mandible and is called the predental. It is thus a synapomorphy of the ornithischian dinosaurs.

There still is no consensus among experts on how exactly the ornithischian dinosaurs developed. It was assumed earlier that *Silesaurus*, the small omnivore from the Middle and Upper Triassic of Poland, which we have already met at the beginning of this book, was maybe one of the first ornithischian dinosaurs. But this is more than unlikely. The fossils of *Silesaurus* come from the Polish Drawno Beds Formation, which formed about 230 million years ago. In addition to skeletons, fossil footprints called *Atreipus* were also found there, and were initially thought to have come from ornithischian dinosaurs. It is now clear that *Silesaurus* left these tracks. The skull of this animal was narrow and, just like that of *Triceratops* many millions of years later, bore a horned beak at the tip of the lower jaw that resembled the predental. However, because the predental is exclusive to ornithischian dinosaurs, this feature was considered evidence of *Silesaurus* belonging to the group of ornithischian dinosaurs. In reality, however, *Silesaurus* did not have a predental. Its horned beak evolved convergently.

Triceratops had another bone in the upper jaw, in front of the premaxilla, which is found exclusively in horned dinosaurs. This so-called rostral was toothless and looked like a parrot's beak. It formed the counterpart to the predental and is a synapomorphy of the horned dinosaurs, which gave them their typical appearance.

How a Cretaceous Landscape is Resurrected

The two *Triceratops* skulls in Frankfurt's Senckenberg Museum are my favorite fossils in this exhibition. I can still vividly remember standing in front of the fossils in amazement as a little boy and my father snapping a photo of my brother and me. I was so excited and fascinated. That was in the early 1980s. These fossils were offered for sale to the museum as early as 1910, and that same year it acquired another important dinosaur fossil: the Trachodon mummy. This is a nearly complete skeleton of a duckbill dinosaur, now named *Edmontosaurus annectens*. It is so unusual because it is still largely covered with its scaly skin. The name *Trachodon* ("mummy") goes back, as it should, to the squabblers Othniel Charles Marsh and Edward Drinker Cope. An associate of Marsh discovered the first known remains of the animal in 1891, and Marsh named it *Claosaurus annectens*. However, because he had not been very thorough, the animal was renamed many times in an attempt to undo erroneous descriptions and duplicate names. It was called *Trachodon* for a while, and was given the name *Edmontosaurus* as recently as 2011. Therefore, *Trachodon* is now more of a nickname rather than a valid scientific name.

My brother (left) and me in front of Sternberg's Triceratops *skulls during my family's first visit to the Senckenberg Museum.*

One of the most talented fossil hunters who worked for both Marsh and Cope—and probably the most successful—was Charles Hazelius Sternberg. He accompanied the two paleontologists on various campaigns into the Cretaceous strata of Wyoming and Montana, and even after the death of the two disputants in the early twentieth century, Sternberg returned there. It is thanks

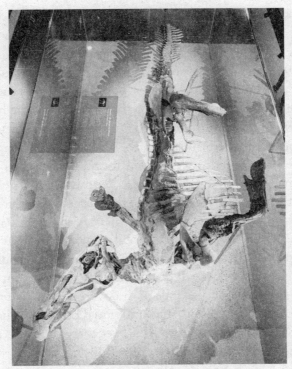

PHOTO BY THE AUTHOR

Charles Sternberg's Edmontosaurus *mummy, now on display at the Senckenberg Museum in Frankfurt, Germany.*

to him that Frankfurt has this mummy. Charles Sternberg, who was sixty years old at the time of its discovery, had previously dug for dinosaurs in the area for over twenty years and had a forty-year career as a fossil collector. He had earned his reputation as an expert fossil collector at a young age when he gathered fossilized plants from near his home in Kansas for paleobotanist Leo Lesquereux of the Smithsonian Museum in Washington. Word of his sleuthing soon spread, and a little later, in 1876, he began searching for dinosaurs full-time for Edward Cope. That was also the year General Custer was crushingly defeated at the Battle of the Little Bighorn by the Sioux, Arapaho, and Cheyenne under their leaders Sitting Bull, Crazy Horse, and Gall in what is now Montana. Not far from this theater of war, in the middle of the Wild West, Cope and Sternberg unearthed spectacular dinosaur finds. Marsh also sought good relations with the Indigenous tribes so that he could dig in their territory. From 1889 to 1894, Sternberg's son Charles, along with paleontologist John Bell Hatcher, searched for vertebrate fossils there on Marsh's behalf and actually found the first remains of the horned dinosaur *Triceratops*—even though he initially mistook its horns as bison horns. Horned dinosaurs are also called ceratopsians, and because their remains were often found in the strata of the Lance Formation, Marsh named these deposits "*Ceratops* beds." Because of the constant weathering at the surface of the site, new fossil discoveries continue to be made there regularly today. According to 2017 and 2018 counts, a total of nearly seven hundred skulls and partial skeletons of horned dinosaurs, one hundred and fifty skeletons of duckbill dinosaurs, and about seventy partial tyrannosaur skeletons have been discovered in the Lance and Hell Creek formations. Teeth and individual bones of dinosaurs are found even more frequently, with over forty thousand documented from these strata, private collections excluded.

Cope and Marsh had both been dead for ten years when Sternberg returned to the region. Starting in 1908, he, together with his sons, George, Charles Mortam, and Levi, worked in the Lance Formation strata in Converse County (now Niobrara County), Wyoming. The county was much larger in the early twentieth century; only later did the eastern part of Converse County become what is now Niobrara County, and only the western part retained its old name. In 1908 and 1909, the family discovered the first *Trachodon* mummy and two *Triceratops* skulls. They sold one of them to the London Museum of Natural History, while the other, along with the duckbill dinosaur, was acquired by the American Museum of Natural History. The Senckenberg mummy came from the southern Schneider Creek area in Niobrara County. In 1910, Charles Mortam, the second-born son of Charles Sr., discovered parts of a dinosaur tail weathering out of the sandstone. Subsequent excavation revealed a complete skeleton with skin impressions. In situ, it measured 17.2 feet, of which the skull accounted for 3.9 feet, the torso 7.9 feet, and the tail 5.4 feet. The recovery of this fossil was the most laborious that the Sternberg family had undertaken up to that time. Charles was determined to secure every fragment of the skin impressions, so the salvaged blocks of rock turned out to be particularly large. The mummy's torso alone weighed about 1.6 metric tons, and the total weight of the fossil was 4.5 metric tons. Since the Sternbergs did not have a block and tackle, the recovery of the fossil blocks could only be done step-by-step by lifting them ever so slightly. It is hard to imagine today how these four men could cope with the tremendous undertaking. The Sternbergs built a ramp of sand and soil and lifted the fossil out of the pit in an immense feat of strength. They used poplar logs as levers and then shoveled sand underneath the blocks of rock. In this way, they were lifted inch by inch to a height of about 3.9 feet before being hoisted onto the horse-drawn carriage and taken to the railroad station at Edgemont in South Dakota, some seventy-five miles away. In total, the excavation took two and a half months.

Charles Sternberg offered to sell the fossil to Fritz Drevermann, then the director of the Senckenberg Museum. Drevermann was able to raise the requested sum through a donation from the industrialist Arthur von Weinberg. However, shortly after the agreement with the German museum director, Sternberg received another offer from the Canadian Museum of Nature in Ottawa. The Canadians offered Charles twice the amount of money for the fossil that he was to receive from the Senckenberg Museum. Sternberg wrote about it in his 1917 memoirs: "I shall never forget the effort I made to induce him to give up the specimen, or take another in its stead. [...] But [the fossil] crossed the Atlantic. The last message I had of it, before this awful war [World War I] cut off all communications, was that the head had been prepared and it was the best of which there was any record."

In the summer of 1910, the Sternbergs also discovered four *Triceratops* skulls, two of which went to the Senckenberg Museum, and a few years later Charles Sternberg found another dinosaur mummy, which he sold to the British Museum of Natural History shortly before the First World War. The fossil was to be brought to Europe by ship. Unfortunately, that ship would be the ocean liner RMS *Lusitania*, which was sunk by a German submarine not far from the Irish coast on May 7, 1915. This terrible tragedy claimed the lives of 1,198 people and the sinking of the *Lusitania* ultimately led to the United States entering the First World War. This is the reason why, to this day, the London Museum does not have a dinosaur mummy and this one lies at the bottom of the sea. More than a hundred years after the ship first sank, however, there's probably nothing left of it.

As interesting as the descriptions of fascinating, novel, and ever bigger dinosaurs may have been back then, nowadays it is no longer enough just to look for the largest fossils. Today, other research topics have become much more important. We want to understand the environment the animals lived in, what they ate, what enemies they had, and what other animals coexisted in the same ecosystem. Scientists want to find out what climate and vegetation prevailed at that time, and they want to learn about the ways of life and cause of death of the dinosaurs. That's why, in 2019, researchers from the Senckenberg Museum traveled once again to the original site in Niobrara County to reconstruct and understand the ecosystem of *Edmontosaurus*. In a collaboration with the Wyoming Dinosaur Center Foundation and supported by the Lipoid Foundation, the Senckenberg team explored the strata of the Lance Formation and recovered an approximately 215-square-foot bone field in July 2019. I had the honor and the great pleasure of accompanying the excavation for *National Geographic Germany* to write four articles about it. For me, it was a pleasure to be part of an excavation for the Senckenberg Museum so many years after my first visit to Frankfurt—even more so in an area where Barnum Brown had found remains of *Tyrannosaurus* and Charles Hazelius Sternberg of *Triceratops*. So, I took a flight from Frankfurt to Denver and rented a four-wheel drive SUV to take me to the quiet town of Lusk. It is located in far eastern Wyoming and has about fifteen hundred inhabitants.

That's 60 percent of the total population of Niobrara County, the most sparsely populated county in Wyoming. Other than a bank, a supermarket, a truck stop, and two liquor stores, there is not much in this town, but it is located in a geologically interesting basin. All around, the land is mostly flat, and extensive pastures stretch out for the many herds of cattle. Here and there you see pumping stations that extract petroleum from rock strata a mile deep. It was here that Sternberg and his sons discovered the Senckenberg dinosaur mummy, even before the county was established in 1911. The rocks exposed north of Lusk are part of the Lance Formation, named for the village of Lance Creek, and contain one of the best-known Late Cretaceous dinosaur faunas. Anyone even remotely interested in dinosaurs knows the names of the animals that have been excavated here: *Triceratops, Edmontosaurus, Ankylosaurus,* and, of course, *Tyrannosaurus.*

The site we drove to was rediscovered back in the 1970s. It is located on the grounds of a private ranch about fifty miles north of Lusk. To get there, you exit US Highway 18 at Mule Creek Junction and end up on a gravel road. Wild sunflowers lined our path, an invasive plant species not originally endemic to the area. The ride was repeatedly disrupted by cattle carelessly crossing the gravel road while pronghorn and mule deer grazed in the distance. With the exception of a small deciduous forest near a river, the area was treeless for many miles. Just before the gravel road ended after ten miles, it took us past a colony of prairie dogs that watched each passing car suspiciously and called out warnings to each other before disappearing into their burrows. From here, the only way to get farther was with an all-terrain vehicle with plenty of ground clearance. Although the site had been known for a long time, it was only then that a scientific excavation was first carried out, which was certainly related to the fact that such a research project was expensive and involved considerable effort. In addition, the site was simply too remote for regular excavations. In the middle of nowhere on the American prairie, some forty tons of ancient bones and rocks had to be lifted, cut up, loaded onto a container, and shipped—a

logistical challenge in a class of its own. Fortunately, the Senckenberg team had support at the time from the Wyoming Dinosaur Center, which had leased the outcrop.

Philipe Havlik, curator at the Senckenberg Museum in Frankfurt, was in charge of the excavation. For him, the site was important because he and his science team wanted to understand what
had happened there in the Late Cretaceous. Today, we know that
during this time period, there were particularly high carbon dioxide concentrations in the atmosphere. In order to find out how
this phenomenon affected the vegetation at that time, a precise
profile of the rock sequence was needed. This would provide us
with information about the type of deposition and, in conjunction
with further lab work, enable us to surmise the era's climate. The
scientists at the Senckenberg Museum were finally able not only to
identify numerous plant and animal fossils, but also to reconstruct
the entire ecosystem in which *Edmontosaurus* once lived. However,
elaborate tooth enamel studies that provide clues to the chemical composition of the atmosphere at the time, just like studies of
plant pollen, can only be done under ideal conditions in a scientific
laboratory. Even small fossils such as mammal teeth, fish scales, or
plant seeds, which are often overlooked in the field, can only be
spotted and quantified in the laboratory. Hence, the question was
how to transfer an entire bone bed to Frankfurt. Philipe Havlik
had an answer: instead of recovering individual bones as Sternberg
once did, he decided on a recovery en block. We were to cut out
all the fossils, along with the surrounding matrix, as one big chunk
of rock. This concept was not new, but to recover a block of 215
square feet was unique and presented a great challenge even to
the experienced Senckenberg crew. The paleontologist Manuela
Aiglstorfer from Mainz, who accompanied our excavation in the
hopes of finding many fossil mammal remains, explained to me
that at first the area had to be uncovered and several feet of overburden had to be removed with the excavator in order to reach the
fossil-bearing layers. In doing so, it was important not to damage
bones that protruded from the rock. Bones that were visible on
the surface were measured and then either recovered directly or

plastered in place. Long steel bolts were then driven into the rock layers of the fossil deposit, and ropes were tied to them to create an excavation grid. This grid helped to mark the exact position of the bones removed, so that they could be reassigned later, in the lab. A grid was then drawn with a spray can and cuboids with an edge length of 3.3 feet were cut out piece by piece. A chain saw with a diamond-tipped saw blade was used for this work, cooled with water from a large tank. Cutting such a bone bed into cubes with a saw, when the bone density is so high, one might risk cutting the bones. But these cuts are rather small compared to the total area and size of the bones, and one can reassemble the cut bones relatively easily in the laboratory. After sawing, the cut surfaces of the cuboids were reinforced with reinforcing fabric soaked in resin to prevent the sediment from crumbling out. Once a cuboid was suitably prepared, it was removed by a telehandler, a type of mini-excavator. Steel plates were attached to the lifting fork, the front edge whetted and its surface greased with lubricating oil. This allowed the telehandler to use sheer force to push the steel plate into the in situ rock beneath the cuboid and dislodge it from the subsoil. Then the oversized "cake lifter" was raised and the remaining surfaces of the rock were also sealed with resin. Now the cuboid only had to be shuttered with plywood sheets and lashed down with ratchet straps. Finally, it was loaded onto a pallet and into an overseas container. Despite all the auxiliary tools, a lot of manpower was needed to handle such a project. That is why Philipe Havlik and Manuela Aiglstorfer were supported by a team of eleven from France.

As soon as we arrived there, it became clear to us that the effort would be worth it, because individual bones of the Edmontosaurs were already sticking out on the surface of the rock. One could see ribs, vertebrae, humeri, and thigh bones with the naked eye, but also more delicate elements such as lower jaws and even finger bones. The size of the bones left no doubt that these were the remains of dinosaurs. Furthermore, the hooves and teeth were indicative of several duckbill dinosaurs. The way they were deposited revealed that the bones must have been moved after death and possibly washed away in a flood, as many different bones from different individuals were crisscrossed. Interestingly, however, there was little to no abrasion at the tips of the bones. This suggested a short transport during which the bones had not rubbed against each other. The exact number of animals buried here could no longer be determined with absolute certainty. In such a situation, however, one can use some simple tools to determine a minimum number of individuals. If, for example, five left femora are found, one can be certain that there were at least five individuals, since such a bone occurs only once per animal. If there are several identical bones of different length, this may speak for animals of different ontogenetic stages. It was clear that the latter was the case with the dinosaurs in our bone bed, and also that it was a herd of a single species.

PHOTO BY THE AUTHOR

Close-up of the bone bed in Niobrara County, Wyoming, showing a vertebra (front) and a scapula (back) of Edmontosaurus.

Philipe was pleased not only with the dinosaurs, but also with smaller finds such as soft-shelled turtle shell fragments, the first mammal tooth of the dig, and another hand-sized piece of rock that showed a deciduous tree leaf next to a conifer branch. Such a fossil would never be seen in the Jurassic strata since deciduous trees did not appear until the Cretaceous period. Layers of leaves or charcoal and individual tree trunks were also found in the bone bed, repeatedly. There was much to suggest that there were mixed forests in this region of the Midwest of North America at that time, indicating to us the prevailing climate at the end of the Cretaceous.

The tension in the field was great because we naturally wanted to get the bone bed to Frankfurt in one piece. Our schedule turned out to be tight and the working conditions were difficult, because all the equipment and tools had to be laboriously transported to the excavation site. In addition, there was no workshop for miles around and no electricity, water, or sanitary facilities on-site—only the burning sun and a lack of shade. The extreme conditions at the excavation site were not only a challenge for the research team, but also put a strain on the equipment. Each block that had to be cut weighed about 1.5 tons, and the total weight of the removed sediment was thirty to forty tons. The chain saw turned out to be the weak point of the project. There were repeated delays due to its defective metal blade and chain. Another issue was the crumbling rock, which broke off at the edges of the blocks despite the resin and could have affected the stability of the whole block during transport. This was mitigated by superglue, of which the team used up dozens of bottles during those days. The coat of resin around the blocks was also a source of danger because, as soon as the toxic, corrosive compound was mixed, it generated intense heat that could have caused burns or even caught fire. Therefore, protective gloves were necessary. Once the resin had been applied and dried, the wooden sheeting was put in place and fastened with the ratchet straps. Then the cavities were filled with polyurethane foam to prevent movement within the block during transport. Whenever a damaged saw blade needed to be replaced, I used the time to search for more fossils on a nearby hill. Philipe told me that he had once found small fragments of a *Triceratops* frill there, and of course I wanted to see for myself if I might discover more pieces of my favorite dinosaur. But since I didn't have the right tools with me, my search was limited to the terrain surface. At least I also found a few frill fragments.

Overall, this excavation was quite impressive. The rock blocks were loaded into a large overseas container with the telehandler. This mini-excavator was just as wide as the inside of the container and was able to push one pallet at a time deep inside. The overseas container was then loaded onto a large truck. To this day, it is still a mystery to me how the truck driver managed to get so close to the excavation site with his huge truck. Even with our much smaller off-road vehicle, we had to carefully drive around holes and large boulders on the gravel road so as not to damage the underbody of the truck. The big truck drove the huge chunk of dinosaur graveyard two thousand miles to the East Coast. There, the fossil bones landed on a container ship and, after millions of years underground, sailed across the vast ocean toward Frankfurt. Later, during the preparation of the blocks, a few predatory dinosaur teeth were actually found—and they were from a *Tyrannosaurus*!

CHAPTER 10

Tyrannosaurus—The Measure of All Things!

The Power of the Brand: Why Tyrannosaurus *Is So Popular*

Every child today knows *Tyrannosaurus rex*. It is the most popular and well-known dinosaur, and is renowned not only in vertebrate paleontology, but also in the hearts of many children. It is a media phenomenon. No other dinosaur has been immortalized so many times in film and television. The scene in *Jurassic Park* in which the muffled stomp of the *Tyrannosaurus* can be heard for the first time, causing the water in the glass to shake even before the giant is seen, is one of the most iconic and spine-chilling moments in movie history. Not least because of the enormous success of *Jurassic Park*, this animal has become a real merchandising powerhouse. No other dinosaur is featured more often on T-shirts and bedsheets. It appears on dinosaur puzzles, on Twitter, YouTube, and TikTok, in memes, cartoons, and comics, and is certainly the most common dinosaur game character (from *Tomb Raider* to *World of Warcraft*, to *Dino Crisis*, *Dinosaur Hunter*, and many more). It is in every bag of dino nuggets, adorns every child's birthday cakes. But what makes *Tyrannosaurus* so unique? What makes it so popular? A lot of it has to do with its name: *Tyrannosaurus rex*! It means "the king of the tyrant lizards," and that's an apt name for the undisputed ruler of the Cretaceous period. *Tyrannosaurus*, however, almost had a completely different name. The first isolated teeth of *Tyrannosaurus* were found in Colorado as early as 1874, but at that time they could not be assigned to any animal. In the 1890s, isolated bones were discovered in eastern Wyoming, and in 1892, Edward Drinker Cope described two vertebrae of a *Tyrannosaurus*, but he mistook them for the vertebrae of a horned dinosaur. He named the animal to which they belonged *Manospondylus gigas*, meaning "large porous vertebrae." If these vertebrae had been better preserved, so that the dinosaur could have been distinguished from other carnivores by their characteristics, then *Tyrannosaurus* would be called *Manospondylus* today. I doubt that this name would have had a similar appeal. In 1900, the American paleontologist Barnum Brown found the first partial skeleton of a giant carnivore in eastern Wyoming, where Edward Cope had also already discovered vertebrae of *Manospondylus* and the first complete *Triceratops*. Two years later, Barnum Brown discovered another partial skeleton of *Tyrannosaurus* in the Hell Creek Formation in Montana. Brown wrote at that time that he had never seen anything like it from the Cretaceous period. Henry Fairfield Osborn, who in the meantime had been named president of the American Museum of Natural History in New York, assigned two names for both animals in a 1905 publication. He called the find from Wyoming *Tyrannosaurus rex* and the animal from Montana *Dynamosaurus imperiosus*, which means "imperial power lizard." A year later, however, he realized that both finds were of the same species, and fortunately decided on the impressive name *Tyrannosaurus*. The lower jaw of the animal he had first named *Dynamosaurus* can still be seen today at the Natural History Museum in London. Since Osborn first published both animal names in the same publication, he could just as easily have chosen *Dynamosaurus*. However, *Dynamosaurus* sounds half as exciting as *Tyrannosaurus*, so we can be glad that he chose the latter name, which we still know and love today! A total of just over thirty complete or at least nearly complete skeletons are currently known. Two of the most famous tyrannosaurs were discovered in the 1990s: Stan, long on display at the Black Hills Institute in South Dakota, and Sue, which can still be admired today at the Field Museum in Chicago. Stan and Sue take their nicknames from their respective discoverers, Stan Sacrison and Sue Hendrickson.

One of the most beautiful skeletons of a tyrannosaur is the so-called Tristan Otto. It was discovered in Montana in 2012 and has the best-preserved tyrannosaur skull in the world. Tristan was on display at the Natural History Museum in Berlin from 2015 to 2020, before being transferred to Copenhagen. It has been back in Berlin since 2022. What makes Tristan special is its black fossil bones. Minerals give the bones a silky sheen. In Leiden, Holland, another tyrannosaur skeleton has been on display since 2016 that is one of the largest and most complete. It also comes from Montana and is nicknamed Trix. From 2017 to 2019, it was the main attraction of a traveling exhibition visiting various places in Europe. I saw Trix in 2018 at a convention in Paris.

A much smaller specimen of a tyrannosaurid was discovered in 1946. American paleontologist Robert Bakker and his team gave it the new species name *Nanotryrannus lancensis* in 1988, which means "the dwarf tyrant from the Lance Formation." It's a curious name, since the remains actually came from the Hell Creek Formation. There is disagreement among experts as to whether this is really a new species or just a young *Tyrannosaurus rex*. Recent findings suggest that it was indeed a juvenile of *T. rex*, but the relatively long arms and its divergent dental formula cast doubt on this. Why it is nevertheless plausible, and what allometry and variability have to do with it, I will come to that later.

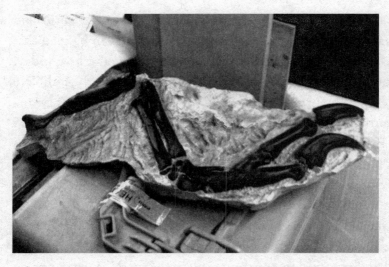

PHOTO BY THE AUTHOR

A cast of the "large" arms of the juvenile "Nanotyrannus" at the Black Hills Institute, South Dakota.

One of the most spectacular finds in my opinion was made in Montana in 2006. A *Tyrannosaurus* and a *Triceratops* were found together, united in a death match in one big block. This can be seen as direct evidence that *Tyrannosaurus* hunted *Triceratops*. The fossil was privately owned for fourteen years until a museum in North Carolina bought it. It was finally put on display for the public in 2022, and I can't wait to see it with my own eyes. The fact that people flock to special exhibits around the world just to see a real skeleton of these animals shows their unwavering popularity. But it is not only the public that is interested in *Tyrannosaurus*. Few other dinosaurs are as well studied as it is, and in the last five years alone, paleontologists have discovered more about this animal than ever before. Today, we understand its biology much better. We know why there were no other larger predators in ecosystems in which *Tyrannosaurus* occurred, we know why its arms were so small, and whether or not it was feathered. Nevertheless, much is still unexplored, and many mysteries surrounding the king of dinosaurs have yet to be solved.

A Record Sum for the King

When I was on the dig in the Lance Formation in Wyoming in the summer of 2019, I met Pete Larson, president of the Black Hills Institute, a dinosaur museum in Hill City, South Dakota. Pete is an extraordinary personality and one of the most important *Tyrannosaurus* discoverers. It is definitely worth googling his name sometime or watching the movie *Dinosaur 13*, which explores the discovery of the tyrannosaur Sue. Pete invited me, the photographer, and the cameraman from the excavation to visit him in South Dakota. I really wanted to see his museum because it had Stan, one of the largest and most complete tyrannosaurs in the world, on display. The animal is extremely impressive, and I definitely wanted to see it at the Black Hills Institute before it was sold, as Pete had told us it would be. Pete and his brother were in litigation at the time, and as a result of the settlement, Pete would keep the museum while his brother would get the proceeds from Stan's sale. Finally, on October 6, 2020, the *T. rex* was sold to an unknown bidder for the incredible sum of 31.8 million dollars. This was a record sum that had never been collected for a fossil before and exceeded the wildest estimates many times over. The previous record holder at auction had also been a tyrannosaur, the *T. rex* nicknamed Sue from the Natural History Museum in Chicago. Its skeleton had already earned its previous owner a whopping 8.36 million dollars in 1992. But as great as the enthusiasm for this high sum was, the experts were shocked. No one knew who the mysterious bidder was, and it was feared that a private collector or investor might withdraw the fossil from public view and prevent future research on the animal. From October 2020 until the end of March 2022, the fossil remained missing, and speculation about the mysterious owner and the animal's whereabouts ran rampant. When a tyrannosaur skull was seen in the background during a television interview with actor and wrestler Dwayne "The Rock" Johnson in January 2022, rumors spread that he might be the one who had purchased the dinosaur at the auction. However, it was only a replica of Stan's head. Two months later, it became known that the future Museum of Natural History in Abu Dhabi had bought the fossil at auction. Many paleontologists expressed relief that a museum had acquired the animal and welcomed the fact that it would now be on display in a part of the world where people rarely had access to fossils and dinosaurs in particular.

PHOTO BY THE AUTHOR

Skull of the large Tyrannosaurus *"Stan" at the Black Hills Institute, South Dakota, before it was auctioned off in 2020.*

How Many *Tyrannosaurs* Were There?

We know *Tyrannosaurus rex* only from the Midwest of North America. Close relatives of the king of dinosaurs, such as *Tarbosaurus*, also lived in Mongolia or *Zhuchengtyrannus* in China. However, the genus *Tyrannosaurus* lived in a spatially limited area. Although *T. rex* is so well-known, it lived for only about

2 million years. How many individuals were there during this time? This question cannot be answered reliably. Nevertheless, colleagues at the University of Berkeley in California have shown with an amazing experiment how one can get a rough idea. They found an approach to this question with the help of the Monte Carlo simulation, a method from probability theory and mathematical statistics (stochastics). The Monte Carlo simulation attempts to solve analytically unsolvable problems numerically. It involves taking samples of a distribution that are determined by random experiments. To find out the abundance of tyrannosaurs, a large number of variables must first be determined: In what time period did *T. rex* live? How large was the area where tyrannosaurs occurred, and what was the population density? How many offspring did a tyrannosaur produce, and how many of these survived to sexual maturity? How old did these animals live on average, and how fast did they grow? Did they have an active metabolism, or were they ectothermic? How heavy was an adult animal? What trophic level did the tyrannosaur occupy, and was there perhaps a change in trophic levels at different stages of the animal's growth? How many generations of tyrannosaurs existed, and what is the absolute number of skeletons that have appeared in the fossil record to date? Then all this data was compared with data from present-day ecosystems, and the results are mind-boggling. The Berkeley colleagues chose *Tyrannosaurus*, of all dinosaurs, for this experiment because it is the best-studied dinosaur, the time interval of its existence was short and the number of individuals small since it was at the top of the food chain. Compared to small carnivores or herbivores, the top predators in an ecosystem are very rare. But what do the numbers say?

Tyrannosaurus roamed the Earth for about 1.2 to 3.6 million years before the asteroid hit. The mean value is 2.4 million years. The Berkeley team calculated the duration of a generation based on the proportion of sexually mature individuals, their average number of offspring, their survival rate, and their maximum lifespan. They arrived at a time span of about nineteen years and thus found that about 127,000 generations of tyrannosaurs must have existed.

The study puts the area where the tyrannosaurs lived at 888,000 square miles. They estimate the average body weight of a sexually mature animal was roughly 11,500 pounds. That's about twenty-seven times the average weight of a lion and about twenty times that of a tiger. Because *Tyrannosaurus* was so heavy and had, like modern mammals, an active metabolism, its energy requirements were quite a bit higher, which is why there were far fewer individuals in an area relative to modern mammalian predators. This is due to the fact that vegetation can only support a certain number of herbivores, and the number of predators depends on the abundance of available prey. Since an adult *Tyrannosaurus* cannot hunt small animals and relies on large herbivores, this limits its abundance. The researchers estimate that the population density of tyrannosaurs was only 16 percent of that of tigers and only 7 percent of that of lions. So, on a safari at the end of the Cretaceous period, you would have been much less likely to see an adult tyrannosaur than you would be to see a tiger or a lion today. It would have been difficult to get a good snapshot of one. But you probably would have encountered many more babies and juvenile animals—though if there's one thing we've learned from the Jurassic Park movies, it's that you wouldn't want to run into a dinosaur in the wild anyway. In absolute numbers, the Berkeley colleagues estimate that the population density of *T. rex* ranged from 0.00058 to 0.14 sexually mature individuals per square kilometer. With a mean of about 0.0091 individuals per square kilometer, there were thus about twenty thousand to twenty-one thousand adults living in the total area at any one time. Calculating the numbers, we find that a total of about 2.66 billion adult tyrannosaurs lived. If one includes hatchlings and young animals, it was certainly ten times more. But if you are familiar with statistics and do the math yourself, you will find that the margin of error is very high. If we follow the same calculation example, but assume the smallest value (1.2 million years) for the reign of the tyrannosaurs and only a value of 0.00058 for the frequency of individuals per square kilometer, then there were only about 84 million adult animals over the entire period of their existence. If instead we assume 3.6 million years and a frequency of individuals of 0.14 per square kilometer, then there were an incredible 61 billion adult tyrannosaurs. Yet these are not the only variables that yield imprecision. If we change, for example, the area size, the duration of a generation, the average age of sexual maturity, the number of offspring, or the average weight, we obtain completely different values.

Why *Tyrannosaurus* Was Not a Scavenger

Every once in a while, some scientist or another will argue that *Tyrannosaurus* may have been a scavenger. A common argument in this context is that its tiny forearms were useless for hunting prey. However, paleontologists Kenneth Carpenter and Matt Smith clearly rejected this claim through a detailed scientific description of *Tyrannosaurus*'s forearms. Since more and more evidence has been found of this animal's very active metabolism, its rapid growth rate, and its incredible bite force, the voices have quieted, although the scavenger argument still comes up occasionally. But even in this case, mathematics can help. Indeed, it is absolutely impossible to feed a six-ton tyrannosaur on carrion alone, as a dear colleague of mine, paleontologist Chris Carbone, impressively demonstrated a few years ago. I learned about his computational model in 2010 at a working group meeting in Flensburg, Germany. A year later his article on the subject was published. The calculations are not quite simple, as they require a lot of research beforehand.

First, you have to estimate approximately how many tyrannosaurs there were in an ecosystem, how much prey and how much carrion was in it, and how long the carrion usually remained accessible as food. Then you have to consider how large the area was where predators and prey coexisted. Knowing these variables, the conclusions are clear. In this computational model, too, we must use values based on observations in nature today. And again, we seek an approach to an intractable problem by proceeding analytically and solving the problem numerically. Weighing the pros and cons of scavenging versus hunting is very important if we are to understand the complex relationships in an ecosystem. This is especially true for the interactions between carnivores of different trophic levels.

As one of the largest land-dwelling carnivores ever known, *Tyrannosaurus* is a particularly popular subject of research, and in the more than one hundred years since its discovery, there has been eager debate about its diet. Was *Tyrannosaurus* an active predator or an obligate scavenger—that is, an animal that was mandatorily dependent on carrion for survival? To find an answer to this question, one must first understand that almost all carnivores do not only actively hunt, but also eat carrion when given the opportunity. Who would turn down a free meal that can be obtained without any effort? Sometimes even herbivores ingest carrion to obtain important minerals when the vegetation in their habitat is particularly low in nutrients.

Nevertheless, few animals rely on carrion as a major component of their diet because the occurrence of carcasses is unpredictable and no strategy can be found for finding carrion at short intervals, regularly, and in sufficient quantity. In all contemporary ecosystems, large carnivorous mammals hunt primarily large vertebrates, and so it is more than likely that a giant carnivore such as *T. rex* also hunted because, with its enormous weight and massive energy requirements, it simply would not have been able to compete with smaller predators as an obligate scavenger.

To illustrate this, we need to look at the environmental conditions of the Late Cretaceous of North America and the size distribution of co-occurring herbivorous dinosaurs, as well as the abundance of competing small carnivores. In the fauna of the Late Upper Cretaceous of North America, surprisingly, we find almost no medium-sized carnivores. There were either small nimble predators with a body mass of less than fifty kilograms or huge tyrannosaurs weighing up to eight tons. If we look at herbivores, the distribution was much more balanced. There were also very large and heavy representatives such as the horned dinosaur *Triceratops*, which weighed up to twelve tons, or the duckbill dinosaur *Edmontosaurus*, with a body mass of up to nine tons. At the same time, however, there were also much smaller animals.

Triceratops and *Edmontosaurus* are the most common dinosaurs found in the Hell Creek and Lance formations. Club-tail dinosaurs like the well-known *Ankylosaurus* are much rarer there; they could weigh up to eight tons. Most tyrannosaurs were found in the Hell Creek and Lance formations of Montana, Wyoming, and South Dakota. But in recent years, remains of these giant animals have also been discovered in New Mexico and Utah. Other herbivores lived there—the largest among them and one of the last long-necked dinosaurs of North America was *Alamosaurus*. Mass estimates of these gigantic dinosaurs vary greatly. A 2020 publication cited about thirty-eight metric tons. Besides these giants, there were also many smaller herbivores, and most animals of this time were nestlings and juveniles anyway. They were much more common than adults because dinosaurs laid a lot of eggs, and their mortality rate was very high. Therefore, my colleagues assumed that almost 50 percent of the herbivores weighed only between fifty-five and eighty-five kilograms.

As mentioned earlier, only about twenty-one thousand tyrannosaurs occurred at any given time over a vast area of 888,000 square miles. Based on the expected frequency of carrion occurrence and the fact that carcasses would have been quickly tracked down and consumed by smaller theropods even before a *Tyrannosaurus* found it, a primary diet of carrion seems very unlikely. Larger carcasses would have been very rare and highly contested, making them an unreliable food source. My colleagues estimated the potential carcass-discovery rate of smaller theropods to be fourteen to sixty times that of an adult *T. rex*. These results suggested that *T. rex* was unable to survive as an obligate scavenger and compete with small predators. Thus, the prey it hunted was primarily large vertebrates, similar to many large carnivorous mammals in modern ecosystems. But what numbers and considerations go into this computational model?

First, we need to estimate how much carrion there was in the first place. Today, a productive modern ecosystem like the Serengeti, in Africa, produces twenty-five pounds of carcasses per square mile per day. If we assume that carcasses were similarly abundant in Mesozoic terrestrial ecosystems, there would have been a substantial number of dinosaur carcasses in North America during the Late Cretaceous—enough to support a single adult *T. rex*. However, if we estimate the distribution of carcasses of different sizes in terms of the frequency they occur, it quickly becomes clear that finding these carcasses would certainly have been a problem. Under the simplifying assumption that all size classes of carcasses occur with the same frequency as the corresponding live animals, we can easily calculate an expected distribution. For this, we draw on the Serengeti production rate and divide that rate among the size classes of herbivorous dinosaurs. Then we calculate how many animals are in each class. If each carcass existed for seven days before ultimately being consumed, most carcasses would be very widely scattered indeed: on average, there would be only a single seventy-five-kilogram carcass within a seventeen-square-kilometer radius, only a single seven-hundred-kilogram carcass within one hundred and sixty square kilometers, and only a single five-ton carcass within one thousand square kilometers. Carrion from a twenty-five-ton herbivore would have occurred on average only once in a five-thousand-square-kilometer radius. Thus, it is clear that these carcasses were difficult, if not impossible, for individual scavengers to locate without traveling very great distances. One animal alone would never have managed to cover such distances and, more crucially, the much more common smaller predators would have long discovered and consumed the carcass in the meantime. Although Chris and his colleagues base their calculations on a much lower *T. rex* population—the authors calculate only about one third of the previously mentioned twenty-one thousand tyrannosaurs (i.e., about sixty-six hundred to seven thousand animals) in the total area—even with higher numbers the chances of finding fresh carrion would have been almost hopeless.

One reason I talk about this, now quite old, article is that the variables have recently regained relevance in the context of why there were no medium-sized hunters in *Tyrannosaurus*-dominated ecosystems. From today's perspective, this unnatural frequency distribution of very large and very small carnivores in an ecosystem where medium-sized predators are completely absent was already noticed by paleontologists early on. For a long time, scientists have tried to understand this phenomenon, but they have not had a good explanation. We cannot see this evolution in ecosystems from the Jurassic. On the contrary, in some of the most famous sites for Jurassic dinosaurs, such as the Morrison Formation in the Midwest of the United States, which we have already learned about, and the Lourinhã Formation in Portugal, carnivores are distributed among all size groups, and lightweight and heavy animals are similarly abundant. This is still true for some Lower Cretaceous ecosystems, such as those of the Wessex Formation in England and the Sao Khua Formation in Thailand.

Particularly fascinating in this context is the Huincul Formation from the Upper Cretaceous of Argentina. There, besides huge carcharodontosaurs that, like *Tyrannosaurus*, weighed more than five tons, middleweight classes are also represented. Only in the Late Cretaceous, as soon as tyrannosaurs became the dominant species, there was this very clear separation between small predators and very large tyrannosaurs, and medium-sized predators no longer occurred. Therein may lie the secret as to why the large dinosaurs were unable to recover from the impact at the end of the Cretaceous.

Different Prey for Father and Son

In a 2011 article, paleontologist Chris Carbone compared different ecosystems of the Late Upper Cretaceous of North America. Various predatory dinosaurs lived in these areas. In his study, it is immediately apparent that on the one hand there were very small predators, for example the dromeosaurs with a body weight of thirty-five pounds, to the one-hundred-and-forty-pound oviraptors, and then there was a huge size gap so that the next largest carnivores weighed around twenty-four hundred pounds. This was more than seventeen times the body weight of the smaller animals. After these predators, there was another huge gap, as the adult tyrannosaurs weighed an average of close to twelve thousand pounds, almost six times as much. We don't know any ecosystems today with such a size distribution. But the authors of the 2011 article did not know that the twenty-four-hundred-pound predatory dinosaur (it was *Nanotyrannus*) was not a separate genus at all, but merely a juvenile tyrannosaur! This means that the actual jump in mass ranges from one hundred and forty to twelve thousand pounds—an almost eighty-five-fold difference! This makes this uneven distribution all the more astounding.

People initially tried to attribute this phenomenon to the incompleteness of the fossil record, and long assumed that medium-sized carnivores simply had not yet been found. The argument has always been a bit specious, but at least plausible, since carnivores are much rarer than herbivores in most ecosystems, and therefore there were certainly fewer predators to be found overall. However, since we have known many of the tyrannosaur sites for one hundred and fifty years, it is clear that this cannot be the reason for the lack of medium-sized species. Moreover, we see the same pattern in dozens of sites in North America and Asia. So what is the reason? The secret lies in the completely different biologies of tyrannosaurs and modern mammalian predators. At the Kruger National Park in South Africa, there are more than fifteen different species of predators. Lions, cheetahs, spotted hyenas and brown hyenas, leopards, African wild dogs and African wild cats, jackals, caracals, servals, among others, can be found there. Moreover, there are African civets, bat-eared foxes, Cape foxes and black-footed cats. And their body masses range from 330 to 550 pounds for lions to only 2.9 to 4.2 pounds for black-footed cats. In between, all weight classes are present: cheetahs and leopards weigh between 46 and 159 pounds; hyenas between 84 and 141 pounds; caracals, jackals, and servals between 14 and 42 pounds; and the small cats and foxes between 2.9 and 14 pounds. Many weight classes are thereby jointly occupied by several species together.

In the dinosaur ecosystems in the Jurassic and up to the early Lower Cretaceous, we still see a greater diversity of species in predatory dinosaurs over fifty kilograms. So medium-sized predatory dinosaurs were also represented there and thus all weight classes were filled. From about 83 million years before our time this changed in East and Central Asia and in western North America (Laramidia). There, tyrannosaurs occupied a monopoly position, and medium-sized adult predatory dinosaurs with a body mass between one hundred and ten and eleven hundred pounds were very rare or even absent entirely, even if they had still been present in these regions before the Campanian. Although predator diversity declined, there was no decrease of prey diversity. The missing food niches in theropod assemblages were occupied by juvenile and subadult tyrannosaur species that were functionally distinct from their adult parents. Herein lies the fundamental difference between the large carnivorous mammals of today and *Tyrannosaurus*: because mammalian young are suckled by their mothers, they indirectly consume the same food as their mothers. Babies of big cats are relatively large and grow quickly. In addition, the cubs are protected by the pride and do not have to hunt for themselves until they are big and strong enough to do so. So, they eat the same food as their parents, even though they are still too small to hunt this prey under their own power. A lion thus always occupies the same ecological niche from birth and is also always positioned on the same trophic level. With tyrannosaurs it was different. Because dinosaurs hatched from eggs that could not grow to any arbitrary size, tyrannosaur hatchlings were tiny compared to their parents. A tyrannosaur egg is about sixteen to eighteen inches tall and a hatchling is about twenty-seven to thirty-two inches long. A baby *T. rex* weighed about as much as a small- to medium-sized dog. The mother of these babies was up to one thousand times heavier. We now know that tyrannosaurs had an active metabolism and grew very quickly. However, because the difference in weight between a hatchling and an adult dinosaur was so great, it still took almost eighteen years for a *T. rex* to become sexually mature. Based on this long time span, we must assume that tyrannosaurs occupied various feeding niches during their lives, some of which were otherwise occupied by medium-sized theropods prior to that time. *Tyrannosaur* juveniles occupied these niches for themselves during their long growth phase because their body proportions changed considerably, and they had to develop new hunting strategies and kill different prey during different life stages. This also means that the novel physique of tyrannosaurs was superior to that of other medium-sized predators; they could not otherwise have outcompeted their rivals. Even a juvenile *Tyrannosaurus*, which had about 10 percent the biting force of an adult animal, was nevertheless stronger and

faster than other predatory species. It thus ensured the evolutionary success of his species and gradually replaced all other carnivores. Some raptors still existed, for example, in the Hell Creek
Formation, but these were extremely rare and do not play a role
in the consideration of faunal assemblages as a whole. One of
these was *Dakotaraptor*, which weighed between 220 and 1,100
pounds. It was about the same size as the raptors in the Jurassic
Park movies and looked similar. In the sixth installment of the
film series, it is even introduced as a separate species, distinct
from the other raptors. In the Hell Creek Formation, this predatory dinosaur may have occupied a special ecological niche not
occupied by the young tyrannosaurs, and its pack hunting behavior probably made coexistence with much larger tyrannosaurs
possible. It may have killed prey that was still too large for juvenile *T. rex*. Besides it, there were some troodontids that weighed
between 110 and 220 pounds. Whether these actually competed
with tyrannosaurs for resources is questionable, since they were
most likely omnivores and did not fulfill much of their dietary
needs with meat. Overall, coexistence in an ecosystem is only
stable if the prey selection of different genera is different. Thus,
we see a decline in species richness among predatory dinosaurs
at the end of the Cretaceous, but not in the number of ecological niches. This has to do not only with the extreme increase
in size of the apex predators, but also with the major changes in
their physique, which were so extreme that experts speak of a
"secondary metamorphosis" of the tyrannosaurs. They changed
their appearance so much that the young animals "functioned"
differently than the adults. Although both belonged to the same
species, they occupied functionally different niches. This is also
a reason why juveniles of *Tyrannosaurus* were not even recognized as such in the beginning but were described as a separate species. Not only did their whole body structure change,
but also their proportions. The initially rather elongated skull,

which looked like that of *Allosaurus*, became increasingly higher in vertical extension, which provided more space for larger and thicker jaw muscles and, throughout ontogeny, ensured that the bite force increased more and more. The ratio of arm length to leg length also changed; in juveniles, the arms were still relatively long but eventually stopped growing while everything else continued to enlarge. This means that a juvenile may have used its arms to catch, injure, or kill prey because its jaws were not yet strong enough to do so. What's more, at some stages of development, the animal's body and tail grew faster than its legs, which is why the adult dinosaurs look beefier and more massive, while the juveniles look slimmer and nimbler. So young animals had proportionally longer arms; their humerus was about 40 percent the length of the femur, whereas that of adults was only 30 percent. Once their jaws became stronger, however, they no longer needed to hold or injure their prey with their hands. Thus, a species occupies different food niches throughout its life. This is called "niche assimilation," and we can observe it in all large dinosaurs.

We observe something similar in the long-necked and duckbill dinosaurs. The Edmontosaurs from the Hell Creek Formation underwent a striking increase in size during ontogeny. Their body structure was functionally different at different stages of their lives. They inhabited a wide variety of habitats during their ontogeny and occupied as many food niches as the smallest grazing artiodactyl, the dwarf antelope, to the largest herbivore, the African elephant. This represents the entire range of ungulates and proboscideans of the Serengeti.

We find no analogy to niche assimilation in tyrannosaurids in the animal kingdom today. Compared to the carnivorous mammals of Kruger National Park, it would be as if a tyrannosaur, in the course of its life, grew from a small black-backed jackal to an animal fifty times heavier than an adult lion! These niche shifts happened abruptly in tyrannosaurids. With increasing size, transitions from one phase to the next were rapid, and transformations in the skeleton and body structure were profound, which in turn changed the way prey were acquired relatively abruptly, too.

In summary, large theropods, unlike carnivorous mammals, occupied multiple niches throughout their lives. This resulted in taxonomic impoverishment, but not a reduction in niche diversity, because juvenile large predators functioned as independent ecological morphospecies. This explains the absence of medium-sized predators in the uppermost Cretaceous. Their niches were filled by the adolescent mega-predators. We can observe this gap in size distribution on different continents, in different habitats, and in different species when the apex predator is from the tyrannosaurid clade. This clearly indicates that this phenomenon is not an artificial effect caused by gaps in the fossil record or a collection bias, but reflects competition for resources in habitats where most niches were filled with juvenile tyrannosaurids. This specialized structure of dinosaurs is related to their extreme size and to the fact that they were egg-laying. Their relatively small eggs hatched into small young that grew very rapidly, resulting in this ontogenetic niche shift. Despite the high mortality rate, the high number of offspring ensured a large juvenile population that morphologically and functionally replaced meso-carnivores. Juveniles were fast and agile and could run with endurance; their parents, on the other hand, had a much higher bite force. This niche separation was necessary because small prey would not have been sufficient for the large tyrannosaurs to meet their energy requirements. In return, the young tyrannosaurs lacked the strength to kill large prey such as adult duckbill or horned dinosaurs, on which the adult tyrannosaurs fed.

I Am the King and Not a Peacock:
Did *T. rex* Have Feathers?

When the first predatory dinosaurs with feathers were discovered in China about twenty years ago, it was a sensation. Over time, more and more finds came to light that showed evidence of plumage in theropods and triggered speculation about whether *T. rex* was covered in feathers also. Since we have now found fossilized integument from tyrannosaurs, we can answer this question.

In biology, the term "integument" refers to the entirety of the dermal layers of animals and humans, including the hair, feathers, spines, scales, and bony plates formed in the skin.

Fossilized remains of *T. rex* and other tyrannosaurids (*Albertosaurus*, *Daspletosaurus*, *Gorgosaurus*, and *Tarbosaurus*) show that large and heavy tyrannosaurids did not have feathers, but instead had a scaly, reptilian-like skin. At least this is true for all adult tyrannosaurids, on the majority of their bodies. We cannot completely rule out the possibility that these dinosaurs wore individual decorative feathers as ornaments. However, there are other *Tyrannosaurus* relatives that were indeed completely covered in feathers, such as *Yutyrannus* from China. Its name is a mixture of Mandarin and Latinized Greek and translates to "feathered tyrant." It is the largest feathered predatory dinosaur disovered to date. Three almost completely preserved specimens of this animal have been found so far. Of these, one is fully adult, one is a subadult, and the third is a juvenile. Although its name suggests it might be, *Yutyrannus* is not a direct relative of *T. rex*, but evolved in a sister group. The fossils of *Yutyrannus* are about 130 million years old and show that gigantism evolved independently (i.e., convergently) at least twice within these sister groups. Here there is a superfamily called Tyrannosauroidea. The higher order is the Coelurosauria, the so-called hollow-tailed lizards. The superfamily Tyrannosauroidea includes two families: the Proceratosauridae, to which the *Yutyrannus* belongs, and the Tyrannosauridae that includes the *Tyrannosaurus rex*. The latter are commonly referred to as tyrannosaurids, and they include all the other known tyrannosaurs mentioned above (*Tyrannosaurus*, *Albertosaurus*, *Daspletosaurus*, *Gorgosaurus*, and *Tarbosaurus*). There is currently no evidence that adult Tyrannosauridae had any plumage. They are uniformly covered with horny scales on the neck, abdomen, hips, and tail. If these animals had feathers, they most likely occurred on the dorsal part. We currently believe that tyrannosaurids lost their abundant pennaceous feathers in the late Lower Cretaceous (the Albian) about 110 to 100 million years ago. Some researchers initially assumed that this loss had to do with the warmer paleoclimate in which the tyrannosaurids lived. However, this is now considered very unlikely. The geographic range of the tyrannosaurids was vast, which is why their representatives lived in extremely diverse climatic conditions. A special example is *Nanuqsaurus*, "the polar bear lizard," because its remains were discovered in the Prince Creek Formation in the extreme north of Alaska. It must have been very cold there during the winter months, and yet there is still no evidence that this animal possessed feathers. Thus, the loss of feathers is apparently not related to the paleoclimate.

According to the first description, *Yutyrannus* was about thirty feet long and weighed roughly three thousand pounds. Other authors assume that the adult animal was only about twenty-five feet long and weighed twenty-four hundred pounds. This may not seem like much of a difference at first glance, but perhaps that is why this animal still had feathers and the larger tyrannosaurids did not. *Albertosaurus*, *Gorgosaurus*, and *Daspletosaurus* weighed about two and a half tons each (some estimates even go up to four tons for the heaviest specimens). *Tarbosaurus* is thought to have weighed as much as about four tons, and *T. rex* averaged perhaps five to six tons, with the very largest animals possibly weighing more than nine tons.

When certain groups of animals undergo an evolution toward gigantism, the ratio of body surface area to body mass changes dramatically. This is because the amount of surface area increases to the second power, but the volume increases to the third power. Such large and heavy animals were indirectly gigantothermic— that is, their bodies maintained a constant body temperature regardless of the outside temperature, because cooling their body through conduction would have taken far too long. Now, if a thick insulating layer had been added, such as a dense plumage, there would have been hardly any heat loss through conduction. Conversely, this would have meant that an animal with a large body mass could quickly overheat. In sauropods, this problem was solved by a complex system of air sacs, and mammals have sweat glands to regulate body temperature. In the case of the heavy tyrannosaurids, the development to gigantism in the course of their evolution had to have inevitably led to the loss of their plumage, so that they did not overheat.

Yutyrannus, which was still fully pennate, apparently had not yet exceeded the threshold at which a full coat would have caused overheating, with its body weight of about one ton. There may have been other reasons that *Yutyrannus* had feathers as opposed to *Tyrannosaurus*—*Yutyrannus* lived in a more densely forested habitat than *Tyrannosaurus* and thus may have needed more protection and thermal insulation—but the deciding factor was the risk of overheating, to which *Tyrannosaurus* was more exposed. In addition, tyrannosaurids had longer legs than their close relatives, suggesting a lifestyle still seen in ratites today. They cover long distances in open terrain and can sometimes run very fast. This behavior is extremely energy consuming and may have increased the risk of over-heating during prolonged chases.

However, because tyrannosaurs are descended from feathered ancestors and almost all hollow-tailed lizards (Coelurosauria) had pennaceous feathers, we now assume that the horny scales of *T. rex* are not comparable to the scales of crocodiles (i.e., not homologous), but that they evolved secondarily from feathers. If anyone had claimed that thirty years ago, they would have been considered simply crazy. But the same is true for the scales on the legs of some modern birds: they also evolved from feathers. This secondary transformation of leg feathers into scales is, in the case of today's birds, an adaptation to wading in water—for example, in storks or herons—or to crouching and balancing on thin branches.

So adult tyrannosaurs were scaly. But what about chicks? Were they feathered or did they already have scales? The fact that adult animals were covered with horny scales does not nec-essarily rule out the possibility that the young still had feath-ers, but that these were lost in the course of their ontogeny. However, such a change from feathers to horny scales is un-precedented in the animal kingdom today, and therefore it will remain ambiguous until we find the first remains of a baby *Ty-rannosaurus* with integument! I would not be surprised if a loss of feathers coincided with a dramatic growth spurt, during a period when the young were transitioning from one ecological niche to another. The small hatchlings may have benefited from the protection of the dense forest and found their prey there, for which they needed feathers, while the juvenile tyrannosaurs went hunting in open terrain and therefore no longer needed them. This scenario is plausible because the juvenile tyranno-saurs, relative to their bodies, had longer legs than the adults and were fast runners.

Short Arms Are Normal but Long Ones Are Strange

Many jokes about *T. rex* allude to its short arms. They were downright tiny compared to its otherwise enormous proportions. Initially, only a humerus and no forearm or digits were found from *T. rex*. Henry Fairfield Osborn, at the first public exhibition of the skeletal reconstruction in 1915, had mounted three-fingered hands on it, as had been seen in *Allosaurus*. But he should have known better, because the closely related *Gorgosaurus* was already known at that time, and it had two fingers on each hand. And although there was later no doubt that *T. rex* also had only two fingers on each hand, it was only a specimen discovered in 1989, which is now on display in Montana, that provided unequivocal evidence of this, because in this fossil the entire arms had been preserved for the first time. Today it is clear that all tyrannosaurids were two-fingered.

But why were the arms of *T. rex* so short compared to its body size? They measured just around three feet, and some of my colleagues describe them as rudimentary. But we shouldn't make fun of it—muscle attachment scars on the humerus, in fact, show that *Tyrannosaurus*'s biceps were very strongly developed. The bones of their arms had extremely thick cortical bone, showing that the animal could withstand heavy loads. The biceps of an adult *T. rex* were so strong that it could lift four hundred and forty pounds with them. Other muscles in the upper arm, such as the brachialis muscle, worked in concert with the biceps to strengthen elbow flexion. So, *T. rex* would not only have won against any human in arm wrestling—its muscle strength would have been enough to simply rip out a human's entire arm.

While human upper arms can rotate three hundred and sixty degrees at the ball-and-socket joint of the shoulder, tyrannosaurs had very limited range of movement. They could only pivot their arms forty degrees at the shoulder and forty-five degrees at the elbow. Biomechanical analyses of the massive arm bones, extreme muscle strength, and limited range of motion suggest that the short arms evolved so that these dinosaurs could hold on to their prey even in the face of massive opposition. The large muscle attachments on the humerus caught Osborn's attention as early as 1906 and he correctly concluded that these arms must have been very strong. He assumed that they were used to hold a partner during mating. In the 1970s, it was assumed that the arms were important when standing up from a prone position to prevent the animal from sliding forward while pushing with its hind legs. All three considerations are plausible and need not be mutually exclusive. In 2021, however, Professor Kevin Padian of Berkeley argued that the reduction of arms in tyrannosaurids had no special function but was rather a secondary adaptation. According to him, the arms became smaller as the animals' skulls became larger and their jaws stronger. This served to prevent bites and serious injuries when, for example, a pack of tyrannosaurs fought over the same prey. But whether this is tenable remains questionable. In Albertosaurs we see such pack behavior. They lived at least partly together with conspecifics of the same age, but for *Tyrannosaurus* so far there is no evidence for such behavior.

The idea that the arms were used as weapons during hunting and the claws were used to slash prey, however, seems more than questionable to me. The arms are much too short for that and their range of motion is, as mentioned above, much too limited. The claws on their fingers could have inflicted long and deep cuts on the prey, but *Tyrannosaurus* probably could not have targeted vital or vulnerable areas with them.

But if you think that the tyrannosaurs had the smallest forearms relative to the rest of their bodies in the dinosaur kingdom, you are wrong. The arms of *Carnotaurus* and *Abelisaurus* from Argentina or those of *Majungasaurus* from Madagascar are much shorter. The three carnivores belong to a group that we call abelisaurians and are found mainly in the southern hemisphere.

This probably seems strange to us because we don't know any other dinosaurs today except for birds. The extremely short arms seem strange to us, even if, conversely, it is actually the anatomy of modern birds that is special. With them a clear decrease of body size took place with a simultaneous increase of relative arm length. The wings are unusually long for dinosaurs. The evolution of birds toward a small body size and long arms probably made active bird flight possible. Indeed, there is conversely an evolutionary trend of negative forearm allometry in theropods that are not yet birds, where larger species often have relatively short forearms. This contradiction can be explained by the fact that the forearm bones, i.e., the ulna and radius, were disproportionately long early in the body evolution of predatory dinosaurs. Thus, the longer a juvenile lived, the longer its humerus and femur became relative to the forearm, which was accompanied by a significant change in arm function. Because tyrannosaurs were not fully grown until they were about eighteen years old, but their arms stopped growing much earlier, the arms of adult animals appear very short. This was not yet noticeable in *T. rex* teenagers, because a three- or four-year-old *T. rex* already had arms as long as its parents, and they simply stopped growing after that. A large abelisaur or *T. rex* killed its prey with its mouth and did not need strong arms. Consequently, the transition from one food niche to the next also determined arm development. Conversely, small theropods needed arms to catch prey and, in the case of birds, to be able to fly. The negative allometry of the arms is thus a signal for evolutionary developments, which are also reflected in ontogenetic trends. In modern birds, the growth of arms is related to changes in movement and behavior during ontogeny—that is, bird flight. The proportionally longer arms of birds are a ju-

venile trait that adult birds retain. We have also seen the same
phenomenon in the dwarfing of *Europasaurus*, supporting the
notion that the evolution of modern birds was driven by paedomorphosis. Paedomorphosis can be achieved by accelerating
sexual maturity relative to the rest of development, or by delaying physical development relative to reproductive capacity.
That's why birds basically look like baby dinosaurs.

Boy or Girl:
Was Sue Female?

The *T. rex* at the Field Museum is named Sue—but is the animal really female? It's hard to tell. Distinct sexual characteristics in dinosaurs are usually lost during fossilization. But one
particular bone tissue is found only in female birds, and in all
other female dinosaurs—medullary bone tissue. It is a special,
derived tissue that lines the internal medullary cavity of females
just before they lay eggs. Medullary bones in extinct dinosaurs
suggest similar reproductive strategies and serve as an objective
method for their sex differentiation. However, there is a catch:
it only helps us identify females that were sexually mature and
about to lay eggs at the time of their death. This is because the
medullary bone served as a calcium reservoir, which the females
needed for eggshell formation. In Sue, we do not find this bone
tissue, which does not necessarily mean that she was a male. It
may simply mean she hadn't laid eggs shortly before her death.
Medullary bone tissue is known from two other tyrannosaurs,
however. One of these animals is on display at the Museum of
the Rockies in Bozeman, Montana. It is nicknamed *B. rex* and
at the time of its death was about eighteen years old and already
weighed three tons. Of *B. rex* less than 40 percent of the skeleton is preserved, but its skull is nearly complete. The medullary bone tissue came from the medullary cavity of its femur,
from which other spectacular discoveries were made, too. In

fact, contrary to popular belief, this tyrannosaur showed that fossil bones can sometimes contain original cells, blood vessels, and structural tissue still composed of its original proteins. In this case, this was due to the extraordinary deposition and preservation of the animal after its death. The carcass of *B. rex* decayed in a brackish estuarine channel and was buried under sand in an oxygen-rich environment, rapidly solidifying. As a result, no further chemical decay occurred. Much research will certainly be done on this in the coming years.

The other female is housed at the Burpee Museum of Natural History in Rockford, Illinois. It was about fifteen years old at the time of its death, still a teenager. Fossils of juvenile tyrannosaurs are extremely rare, but the Burpee Museum has not one but two of them, both found in the same region of Montana. The more complete specimen of the two teenagers, nicknamed Jane, was slightly smaller and, at about thirteen years old, a bit younger. However, we do not yet know if Jane was a female or even sexually mature. Paradoxically, medullary bone tissue was detected in the larger animal, which bears the male nickname Petey. Jane, by the way, is that specimen that was thought to be the new species *Nanotyrannus lancensis*. However, since it turned out that it was actually just a young *T. rex*, the animal is called *Tyrannosaurus* again.

The King, the Queen, and the Emperor

With all the confusion about gender and dramatic changes in body structure, it is not surprising that some of my colleagues believed that there may have been several different species of *Tyrannosaurus*. The species *Tyrannosaurus rex* is widely known, but researchers have long observed that some animals were more robustly built and some had wider pelvises than others. Some paleontologists interpreted these as gender-specific traits. They argued that the narrow pelvis was indicative of males and the wider one was indicative of females, because a wide pelvis would certainly be advantageous when laying eggs. Others saw the high variability as a sign that there must have been several different species. This hypothesis does not seem to me to be completely far-fetched; after all, as we have already learned, *Tyrannosaurus* existed on Earth for about 1.2 to 3.6 million years. This time span is sufficient for it to have split into several species. A 2022 article states that there was originally one species of *Tyrannosaurus*, which later gave rise to two new ones. The authors noted that many footprints and skeletons of *T. rex* showed considerable variability that had not yet been studied in detail stratigraphically. This meant, they said, that it is not yet known whether the differences might not be related to a different temporal occurrence of the respective finds. In studies of more than three dozen specimens of *T. rex*, the authors found evidence of a remarkable disproportionate variation in tyrannosaurs, which they believed could not be explained by different growth phases or sexual dimorphism alone, but by the fact that various forms occurred in chronological succession and were found in different rock strata of different ages. They cited the robustness of the skeleton and the varying number of small incisor-like teeth in the lower jaw as the most important arguments for their theory. Thus, at the beginning of the tyrannosaur evolution, there was one robust species with two small teeth in the front and much larger teeth farther back in the lower jaw. The robustness of the animals was defined by the circumference of their femora. The original form of *T. rex* is thought to have later split into two new species that had only a single small tooth in front of the larger teeth in the lower jaw. One of these new species still had robust femora and the other one had more slender ones. The two new species then replaced the original species.

When a new species develops from an old one, it is called anagenesis. When an original species has split into two new sister species occurring at the same time, a so-called cladogenesis took place. Something like this can happen when two populations are spatially separated and then adapt differently to their respective habitats.

The first skeleton of a *T. rex* ever found was not at the same time the geologically oldest. It was a more derived form as a result of the anagenesis, and a robust one at that, but bearing only one small tooth in the anterior mandible. However, according to the rules of nomenclature, the specimen which was described first must retain its original name, so the authors gave this species the name *Tyrannosaurus rex*. For the oldest robust form with two small teeth at the tip of the lower jaw, they later gave the name *Tyrannosaurus imperator*, while they renamed the more gracile new species *Tyrannosaurus regina*. I like all of these names very much. *Tyrannosaurus rex*, as we recall, means "king of the tyrant lizards," *Tyrannosaurus imperator* means "emperor of the tyrant lizards," and *Tyrannosaurus regina* means "queen of the tyrant lizards." According to this new naming scheme, Sue would be a *Tyrannosaurus imperator*, and a *Tyrannosaurus regina* can be seen at the National Museum of Natural History in Washington, DC. However, it is uncertain whether these names will catch on in the long run. I found the distinction of these three species based on only two characteristics very questionable, and shortly after the publication of the article in question, many experienced colleagues shared my concerns. There are doubts about the coherence of some arguments. If one postulates such a distinction of different species, one should clearly prove it. Thereby the assumption of a species succession over a period of 2.4 million years is quite plausible. In the case of *Triceratops*, the species *Triceratops horridus* was followed by the species *Triceratops prorsus* in the same period. On the other hand, there are very many skeletons of *Tyrannosaurus* that have hundreds of distinct skeletal features by which its species can be described. To derive three species on the basis of only two features does not seem very convincing to me. But the authors did not succeed in finding any more features among the many that support their hypothesis. In fact, femur circumference varies widely among tyrannosaurs, and other authors have shown that there are not two size clusters with thick femora on one side and slender femora on the other, but that there is a whole range of bones of different thickness, with some occurring somewhere in the middle.

The tooth variation is a groundless argument. With mammals, this characteristic would have perhaps been convincing, since they usually have a strict tooth formula so all animals of the same kind always possess the same number of teeth. Humans, for example—so long as they brush their teeth thoroughly—always have two incisors, one canine, two premolars, and two molars per quadrant. As they get older, they then grow one more molar per quadrant, which we call wisdom teeth. Reptiles, on the other hand, have very variable tooth formulas that allow for much more plasticity. Two animals of the same species may have more or fewer teeth in their jaws. To base different reptile species on different tooth formulas therefore does not seem very reliable to me. That the distinguishing features are also inadequate becomes clear from the example of the New York *Tyrannosaurus*, which is one of the most complete specimens in the world. In this animal, the skull, ribs, pelvis, and the entire spine up to the tip of the tail are almost completely preserved. It is only missing its thighs. Consequently, it would be impossible to assign this animal to a *Tyrannosaurus* species because we do not know the circumference of the femora. It is not surprising that some colleagues refuted the corresponding article only half a year after its appearance. Perhaps someday we will actually be able to determine differences that will make it necessary to divide the genus *Tyrannosaurus* into several species. But so far this is not necessary, and so the name of our favorite dinosaur remains with us for the time being: *Tyrannosaurus rex*!

CHAPTER 11

Movement Captured in the Stone:
What Footprints Tell Us

The German Dinosaur Highway

Much of what we know about dinosaurs, we have learned by closely examining their bones. The size of the bones suggests the overall size of the animals, and the circumferences of their arm and leg bones give us a rough idea of how heavy dinosaurs were. Processes and muscle attachment scars on the bones tell us how strong the muscles were and where they ran across the body. The internal bone structure tells us if the animals had an active metabolism, how fast they grew, and when they were sexually mature. Bone cells allow us to draw conclusions about genome size. Medullary bone helps us determine their gender. From the feathers of enantiornithines found in amber, from the dermal bones of the armored dinosaur *Borealopelta* found in the oil sands of Canada, and from the scales of dinosaur mummies and tyrannosaurs, we get an accurate picture of the dinosaurs' appearance and their biology. As a young boy, I never dreamed that one day we would find out what color dinosaur eggs were and what the cloaca or lungs of these animals looked like. Our picture of the prehistoric lizards is becoming clearer, and it's one of dynamic, agile, and intelligent animals. Today, we know that some dinosaurs displayed brood care behavior and were caring parents. The dinosaurs of my childhood were clumsy, lumbering, and all gray. Today, we see them on Apple TV in the most dazzling rainbow colors and complex food webs. But even though the fossilized bones, scales, and feathers reveal a lot about the life of the dinosaurs, they tell us little about the everyday lives of these animals. They cannot capture a moment in the lives of dinosaurs. Only the so-called trace fossils can do that. They are like a snapshot frozen in time over millions of years. Unlike bones, which are often transported, trace fossils remain exactly where they were produced. Scientists call such fossils, which remain in place, "autochthonous." Other examples of autochthonous fossils are fossilized reefs. If we find dinosaur tracks, we know that at some point in Earth's past, a living animal walked in that exact spot. We can tell whether the animal moved fast or slow, was heavy or light-footed, whether the ground was sandy or muddy, and whether the animal walked alone or as part of a herd. Because sauropod tracks are autochthonous, they provide unambiguous information about the actual habitat of the track maker. The giant tracks of gigantic sauropods occur in sediments from the Late Triassic to Cretaceous periods around the world. In the fossil record, they are the second most common evidence for the long-necked giants. Tracks are important because they reveal anatomical details and locomotion patterns of the track maker that are otherwise difficult to determine.

Some of the best-known tracks are in mudflat deposits of the Paluxy River near Glen Rose in Dinosaur Valley State Park in Texas. More dinosaur tracks were discovered in the early 1980s, in the Kugitangtau Mountains of eastern Turkmenistan at an elevation of five thousand feet—sometimes called the "dinosaur plateau." They are among the longest continuous tracks ever found. New methods in studying the tracks are helping us to better understand dinosaur biology. The principles of soil mechanics inspired a field experiment with an African elephant whose footprints were used to infer the animal's weight. The geometry, depth, and diameter of the footprints, in combination with soil mechanical analyses, were used to calculate the mass as reliably as possible. Subsequently, it was intended to apply this method to dinosaurs as well. The experiment was conducted at the Wuppertal Zoo with a female elephant. The elephant was weighed, and the track was prepared so that deformations in the sand could be measured accurately. The geometry of the footprint was determined by laser scanning and the velocity was determined with digital image correlation. The properties of the soil were determined in advance by experiments in the laboratory. Finally, for the reverse calculation of the elephant's weight, the finite element method was applied. The study showed that footprint geometry, together with theoretical considerations of subsurface loading, is well suited for predicting the weight of the track maker. Because this method can be used to reliably measure the applied additional load on certain substrates and calculate an animal's weight with an error rate of only about 15 percent, it will certainly play a major role in dinosaur mass calculations in the future. But the process is complex because several properties of the footprints and the substrate must be taken into account before reliable results can be obtained for the fossils. In addition, researchers must account for geological processes that may have altered the original shape of the sediment. The method only works reliably in sandstone because sand is not further compressed during its diagenesis to sandstone, whereas in mudstone, measurement accuracy is compromised.

A grain size analysis and a detailed sedimentological investigation are therefore important in advance.

An ideal place to test this method was the dinosaur track site of Münchehagen, a natural monument in a former quarry, where hundreds of sauropod and theropod tracks can be found over an 160,000-square-foot area. There—about 140 million years ago, at the beginning of the Lower Cretaceous—the dinosaurs left their footprints in the soft subsoil of an estuarine delta. These tracks can be traced further in the adjacent quarry.

The Dinosaur Tracks of Lommiswil

Sometimes, tectonic movements of the Earth's crust can cause dinosaur tracks to be exposed vertically, as can be observed in Switzerland, halfway between Basel and Bern, near Lommiswil. The excursion to Frick, where I saw the plateosaurs, had also led me to the track site of Lommiswil. The tracks there were only discovered about thirty years ago, although the nearly vertical rock face has been exposed for eighty years. One would simply not expect the sauropod tracks, which are showcased here, to be found in marine deposits. The tracks are from the Kimmeridgian, which corresponds to the "White Jurassic," and are dated to 157.3 to 152.1 million years before our time. They are thus only slightly older than the tracks from Münchehagen. In the Upper Jurassic, sea level fluctuations apparently led to temporary silting at this site, allowing dinosaurs to enter the otherwise water-covered area. The partially visible mud cracks indicate a period of drying of the water body. They form when the mud contracts as a result of receding water. At that time, the sauropod dinosaurs walked through calcareous mud plains, deposited in tidal pools, which are now known as the "Solothurn Turtle Limestone." Indeed, sea turtle fossil finds are also known from there. The traces of the hind feet of these dinosaurs have a diameter of more than three feet, and those of the front feet are somewhat smaller, rather horseshoe-shaped, and less clearly visible. Thus, the front legs did not sink as deeply into the mud, as the main weight of the animal's body rested on its hind legs. These animals were toe-walkers, and their toes were embedded in a large pad of gristle similar to the shock-absorbing pad of today's elephants. The tracks of the hind feet of the sauropods have a circular to oval shape and no claws are visible. We can therefore assume that the trackers were probably not *Diplodocus* relatives, since three claws on the hind feet are characteristic of this group. Depending on how soft the mud that the sauropods walked on was, this detail might simply not have been preserved, even though the animal may have had claws after all. A better indication is the width of the tracks.

PHOTO BY THE AUTHOR

The steep face with sauropod tracks in Lommiswil, Switzerland.

Although *Diplodocus* is a long-necked dinosaur with one of the longest necks, it was quite delicately built. It had a narrow pelvis and was more like a greyhound among sauropods rather than a bulldog. The narrow hips meant that the legs were cantered under the body, and the animal put one foot right in front of the other while alive. If a line were drawn running between its left and right footprints, it would intersect both footprints. Later in the evolution of long-necked dinosaurs, the titanosaurs, on the other hand, had such massive bodies that they walked with their legs wide apart. The tracks at Lommiswil show such a broad-gauge gait.

There are *broad-gauge trackways* and *narrow-gauge trackways*. These are terms derived from railroads, where there are narrow-gauge and broad-gauge tracks.

Titanosaurs did not yet exist in the Upper Jurassic, but the brachiosaurs also already walked with a broad-gauge gait. Their tracks were not caused by a herd moving together in the same direction, but by individuals that crisscrossed. Therefore, no clear predominant walking direction can be recognized. The animals moved very slowly and were probably searching for food. Calculations showed a walking speed of only two to three miles per hour. Spore findings show that various conifers and fern species were available as food sources for these animals. Conifers were well adapted to the prevailing arid climate, but also to brackish conditions. In addition, the sauropods may have fed on trees with water-storing leaves and herbs that grew on the tidal flats. The tracks usually stand out from the rest of the rock by their lighter coloration because the tracks were subsequently covered and infilled with carbonate plugs. However, as the escarpment is exposed to weathering, the tracks are beginning to fade and will become less and less distinct in the future.

The Track Site of Cal Orck'o in Bolivia

Some of the most impressive dinosaur tracks are located near the city of Sucre in Bolivia. At the Cal Orck'o mountain, a paleontological site, there is a cliff with thousands of footprints. It is about two hundred and sixty feet high, four thousand feet wide, and has an incline of seventy-three degrees. The tracks come from hundreds of different turtles, crocodiles, small lizards, but also from sauropods, theropods, and ankylosaurs from the Upper Cretaceous period. The wall in central Bolivia has been known since 1968, and recent work shows that track sites also exist elsewhere in Bolivia, occurring in several strata from the Campanian to the late Maastrichtian (83.6 to 66 million years ago). They all belong to a mega-track complex that extends from southern Peru across the central Bolivian Andes to Salta Province in northern Argentina.

This is probably the largest assemblage of dinosaur footprints in the world. The main track area is about two hundred and seventy thousand square feet and was mapped in 1998 using heavy mountain-climbing equipment. The track-bearing strata show episodic soil formation, stromatolites, and storm deposits. They formed in calcareous lake deposits that repeatedly dried out in places. Three hundred and thirteen tracks were recorded on nine levels, including those of five different dinosaur species. Tracks of ankylosaurs were the first to be discovered there. The high diversity of tracks clearly shows that the decline of dinosaur species toward the end of the Cretaceous, at least in South America, was not gradual, but that the dinosaurs died out abruptly due to external factors.

CHAPTER 12

Birds—The Last Dinosaurs

Where Is the First Ostrich?

The only dinosaurs that survived the mass extinction at the end of the Cretaceous period are the modern birds. We can divide them into two major groups: the primitive "old jaw" birds (Paleognathae) and the "new jaw" birds (Neognathae). The distinguishing features of these two groups are, among others, the palatal roof on the beak of the birds and their pelvis, where there is an opening between the ischium and the hip bone in the paleognaths. This opening in paleognaths is clearly reminiscent of other non-avian dinosaurs. In neognath birds, the skeleton is already further derived and differs more from the original dinosaur bauplan.

The paleognath birds include, for example, ostriches, cassowaries, emus, and the South American nandus, which I saw in the wild during my excavation trip to Argentina. The extinct moa and the elephant bird belong to this group also. The surprise representative of the group is the kiwi, which is closely related to the cassowary. The only paleognaths that can fly are the tinamous. Like the nandus, they originate from South and Central America and are found in the tropical lowlands. All other birds belong to the subclass of neognaths, which can be further divided into the Galloanserae and the Neoaves. The word *Galloanserae* sounds complicated and hard to remember, but in reality it is just a composition of the two Latin words *gallus*, which means "chicken," and *anser*, the Latin word for "goose." Accordingly, this group includes all galliformes and all waterfowl. The galliformes include pheasants, quail, and guinea fowl, and the waterfowl include geese, ducks, and swans. The name *Neoaves* simply means "the new birds." They include 95 percent of all birds alive today, which is still over ten thousand species. By contrast, the paleognaths, i.e., all ratites and tinamous, together comprise only about sixty extant species. But although their diversity is very minimal compared to the new birds, they occur on three continents and play important roles in their respective ecosystems. Ostriches and their close relatives are quite poor at distinguishing colors. This is an important difference from the neognath birds. I will explain why this is significant when I write later about color vision in dinosaurs and birds. All of the paleognaths lay their eggs in nests on the ground, with the rooster performing many of the brood care tasks and helping the hen hatch the eggs. The eggs of this group are the largest in the animal kingdom, led by those of the ostrich. However, the eggs of the extinct elephant bird were much larger. The content of one such egg was equivalent to the mass of about one hundred and ninety chicken eggs. The egg of the kiwi is much smaller, but this bird lays the largest eggs relative to its body size. A kiwi egg can weigh up to five hundred grams, which is about 30 percent of the mother's body weight. In this case, the egg fills almost the entire abdominal region of the hen. The size of these eggs can be explained by the fact that paleognath chicks are precocial and already very well developed by the time they hatch. Their bodies are more developed than those of altricial chicks even when they have just hatched. As a result, the chicks are also larger and often already have plumage, whereas some altricial chicks are still naked and blind at hatching. Although it can be advantageous to give birth to young that are already well developed and no longer need intensive care from their parents, chicks cannot remain in the egg indefinitely. The size of the eggs of ostriches, moas, and elephant birds is already approaching biological limits. Even dinosaurs that weighed ten or even a hundred times more than ostriches could not lay larger eggs. The egg of the largest moa, *Dinornis*, is about as big as that of a *Tyrannosaurus rex*, although the *Tyrannosaurus* weighed about fifty times more than *Dinornis*.

But why is the *T. rex* egg not larger? Why can eggs not grow to any size? As we all know, the shell of bird eggs is made of calcium carbonate ($CaCO_3$) and is very porous to allow the embryo to breathe air in and for carbon dioxide to escape when it exhales. With a larger egg, the thickness of the shell would also have to increase to keep the egg stable and prevent it from being crushed by its own weight or that of the incubating parents. But the thicker the shell, the less air gets inside, and the less carbon dioxide (CO_2) can escape; ultimately the unhatched chicks would suffocate. In the case of the ostrich, this is exactly what happens during the last phase of growth in the egg: the chick literally loses its breath. And when it then convulses and begins to twitch inside the egg as a result of its impending death from suffocation, and due to the violent spasms and kicks of the chick inside, the thick eggshell finally breaks. Male ostriches in the wild weigh about two hundred and forty pounds and females slightly less. In the first year of life they gain about two hundred and twenty pounds in body mass—so they grow incredibly fast, which is only possible due to their active metabolism. To do this, they need a lot of food, but also a lot of oxygen. However, while the surface of their egg grows according to a quadratic function, the volume inside increases to the third power. So, if the egg grew any larger, the ratio of surface area to volume would change so unfavorably that not enough air could get inside and the chick would suffocate, since the surface of the eggshell is its only access to atmospheric oxygen. Conversely, as the size of the embryo increases, so does the amount of carbon dioxide that accumulates inside the egg. If the shell becomes too thick and carbon dioxide is exhaled regularly at the same time, it can no longer escape quickly enough. The same is true for dinosaurs. Today we know that these prehistoric animals, at least in their growth phase, had an active metabolism comparable to that of birds. They, too, grew very quickly, and they, too, would have run out of air in even larger eggs.

But paleognaths are fascinating not only because of their unusually large eggs. Their feet strongly resemble those of large predatory dinosaurs. Ostriches, nandus, cassowaries, and emus are also much larger and significantly heavier than any birds capable of flight. The giant Kori bustard, weighing up to forty-two pounds, is the heaviest bird capable of flight. But even the nandu, the smallest ratite, already weighs about fifty pounds. Emus weigh up to eighty-two pounds, cassowaries about ninety-seven pounds, and the African ostrich weighs up to two hundred and forty pounds—ten times heavier than a swan. Although ostriches are all vegetarians, some species can be very aggressive and dangerous. There are several documented cases where cassowaries have fatally injured humans with their sharp claws. It is hard to imagine how much worse the *T. rex*'s attacks must have been, considering it weighed two hundred times more.

The similarity of the paleognaths to the large predatory dinosaurs suggests that this group of modern birds arose earlier than the neognaths. Some scientific studies involving genetic data and paleontological values indicated that all modern orders of birds, including the paleognaths, must have originated in the Early Cretaceous and even then began to branch off from the phylogenetic tree of the giant theropods. Molecular data suggest that the paleognaths evolved as early as 110 to 120 million years ago in the Lower Cretaceous, just as the first enantiornithines did. Unfortunately, we have no fossils from that time. The oldest unequivocal evidence for a paleognath in the fossil record comes from strata that formed only after the impact of the asteroid. The oldest member of the group known to date is called *Lithornis celetius* and comes from rocks from the Middle Paleocene of Montana and Wyoming. It is about 61 to 60 million years old. But where are its older relatives?

If paleognaths were formed 110 to 120 million years ago, but the first fossils do not appear until the Paleocene, this leaves a gap of about 50 million years in the fossil record, which is hard to explain. When there is such a large gap between the hypothetical appearance and the first actual fossils of a group, it is called a ghost lineage. We know that these animals should be found somewhere, but we have no trace of them. Perhaps older specimens will indeed be found one day, closing the gap. But maybe the molecular data has led us down the wrong track. In the meantime, there is growing evidence that the diversification and evolution of these animals accelerated after the asteroid impact. It is possible that we are dealing with a phenomenon similar to the rapid diversification of ichthyosaurs at the beginning of the Triassic. Perhaps the speed of molecular evolution was underestimated and perhaps the evolution of birds—just like in ichthyosaurs—accelerated after the impact of the asteroid as a result of adaptation to new habitats. The last common ancestor of all present-day paleognaths could also have arisen just before the end of the Cretaceous, surviving the extinction event and only then evolving from a small form to a large, flightless animal. Unusually high mutation rates would have ensured rapid diversification.

However, the question of the first appearance and origin of the group is not the only unsolved mystery in this context. Today's paleognaths all come from countries in the southern hemisphere that once belonged to the continent of Gondwana.

> The large continent Gondwana consisted of today's
> South America, Africa, India, Australia, and Ant-
> arctica.

Therefore, it was long assumed that this group originated
there. However, molecular data, the fossil record, and phylogenetic studies over the last twenty years refute this hypothesis. Animals such as *Lithornis* also suggest that the evolution to
flightlessness and large body sizes occurred later in their evolution. So, the question arises whether it is possible that these
two traits evolved three times, independently of each other,
in South America, Africa, and New Zealand, at a time when
these landmasses were already far apart. Thus, we still do not
know exactly how, when, and where the present paleognath
diversity arose. The group does appear to have lived before the
Cretaceous–Paleogene boundary (K–Pg), but since we do not
have Cretaceous fossils, we cannot trace the convergent transitions to flightlessness and large body sizes. And while *Lithornis*
is about 60 million years old, we don't even find direct ancestors of modern paleognaths until about 40 million years later.

Moreover, the following thought is keeping me up at night:
paleognaths and neognaths are the only surviving dinosaurs. In
the group of archosaurs, only the modern crocodiles have survived. This means that the closest relatives of the paleognaths
outside the avian lineage are the crocodiles. This split of archosaurs into crocodiles and dinosaurs, however, already occurred
250 million years ago. So, if we compare features in the family
tree or in the genome, it is not surprising that all paleognaths
always end up in the same group. They are different from the
other birds and are light-years away from crocodiles in their
development. But maybe this effect is just related to the fact
that we don't have any dinosaur DNA available yet? Maybe the
paleognaths are not a uniform group at all, and their similar appearance is only a result of convergent evolution? Perhaps the

bioprovince of Australia was already so isolated that cassowaries and kiwis descended from a different coelurosaur than ostriches, nandus, and tinamous. Or maybe it all happened much later. We can only hope that at some point there will be finds that will allow us to reconsider the relationships of paleognaths to neognaths and their dinosaurian ancestors. This would also explain the long ghost lineage, because there is no direct common ancestor. Perhaps ostriches, nandus, and cassowaries split off from coelurosaurs at three different nodes.

Almost Like Easter:
An Egg Thief Who Lays Colorful Eggs

All birds lay hard-shelled eggs. They are also the only terrestrial vertebrates, some of which lay colorful eggs. Crocodiles and turtles are a little different. They also lay eggs, but these are white because their shells do not have color pigments. From whom did birds inherit the ability to lay colored eggs? Were their dinosaurian ancestors also able to do this? Did all dinosaurs lay colored eggs? And why do birds lay colored eggs while crocodiles do not?

That some birds produce colored eggshells is a result of selective pressure. Egg coloration is primarily for camouflage. If they have the same coloration as the vegetation that surrounds them, or if they have speckles to blend in with gravelly substrate, they are less likely to be discovered, and are less likely to be eaten by egg thieves. Egg camouflage is thus an important part of the breeding behavior of open-nesting birds, which protect their unhatched offspring from visually oriented predators in this way. But not all eggs of open-nesting birds are colored. In fact, many birds protect their eggs by continuously incubating, so coloring is not necessary for them. Colored eggs also help birds distinguish their own eggs from those of other species, and it keeps them from mistaking foreign eggs as their own, when a

parasite secretly adds them to the nest, as is the case with cuck-
oos. Turtles, on the other hand, bury their eggs in the sand, and
crocodiles cover them with branches or twigs, so they don't need
camouflage. If the egg of an extinct animal is colored, then we
can conclude that it was not buried, but laid in an open nest.
The color of its egg thus allows us to draw conclusions about the
social behavior of an animal, even if we can no longer observe
it in the wild. This has led dinosaur researchers to investigate
whether dinosaurs laid colorful eggs even before modern birds
emerged from them. Fortunately, there are many sites with dinosaur eggs around the world. The first specimens were discovered in France in 1846. Other finds have come from the United
States, Spain, and China, where entire nests and clutches have
been found containing dozens of eggs.

Jasmina Wiemann, a colleague at the Field Museum, wanted
to know if only the direct ancestors of birds produced colorful
eggshells or perhaps long-necked dinosaurs, duckbill dinosaurs,
or horned dinosaurs as well. She studied dinosaur eggshells from
around the world and used microspectroscopy to identify different pigments responsible for bluish, greenish, and reddish-
brown coloration. She found that color pigments are present
only in the eggshells of theropods, but not in other dinosaurs.
She also found that among carnivorous dinosaurs, only Maniraptora had colored eggs. Carnivores such as *Tyrannosaurus* and
Spinosaurus did not lay colored eggs.

**Maniraptora means "Seizing Hands." These dino-
saurs evolved a crescent-shaped wrist bone that
allowed them greater hand mobility, which was a
prerequisite for bird flight. Feathers also evolved
within the Maniraptora group. In the Mesozoic, these
animals tended to be small- to medium-sized and had
the largest brains among dinosaurs relative to their
body weight. The group also includes modern birds.**

Jasmina Wiemann was able to trace the origin of colored egg-shells back to an oviraptor from the Late Cretaceous of China. Oviraptors are those toothless maniraptors we know as "egg thieves." Yet they were probably very caring parents, hatching their chicks in their nest and nurturing their young. Their nests contain eggshells of the type *Macroolithus yaotunensis*, assigned to the species *Heyuannia huangi*. Some of these eggs contained exceptionally well-preserved embryo remains that can be assigned to *Heyuannia*. They come from three river deposits from eastern and southernmost China. The pigments in these eggshells indicate a blue-green coloration. The colored eggs of *Heyuannia* are thus the oldest in the fossil record. Their pigmentation supports the interpretation of their depositional conditions—because the eggs were colored, they were laid in an open nest and were not buried. The nests of these oviraptors are mostly circular and the eggs in them are arranged in several concentric layers, stacked on top of each other. All the while, the elongated eggs are stuck almost vertically in the sediment, with the pointier side down. In the circle, they are also arranged in pairs, with two eggs always close together and separated from the next pair by sediment. This arrangement, as well as the shell pattern and shell porosity, are clear indications of an open nesting behavior of *Heyuannia*, because buried eggs, as those of turtles, for example, are not neatly, symmetrically arranged.

It is interesting to note that no bird today arranges its eggs in this way. Only in the nests of ratites, which live in breeding colonies, can a similar breeding behavior be observed. Some of these birds lay green ones, like emus, or bluish ones, such as cassowaries. The nest arrangement and blue-green eggs indicate that oviraptors were already engaged in intensive nesting. The colors blue or green are usually not used for camouflage, but as signal colors. They are especially common in birds, whose roosters play a major role

ARMIN SCHMITT

in parental care. So, we can learn and infer a lot about the social behavior of these animals from the color of their eggshells.

Because we have not found colored eggshells in other dinosaur groups, the blue-green eggs of *Heyuannia* also show that colored eggshells did not evolve multiple times, but only once in the Maniraptora group. So, if someone asks you in the future who brings the colorful Easter eggs, the answer should definitely be the Maniraptora!

The Goddess Asteria:
The Mother of All Chickens and Geese

The origin of modern birds remains difficult to find, as shown by the ghost lineage of paleognaths. We cannot trace the earliest evolutionary stages of modern birds because of the incompleteness of the fossil record from the Mesozoic. Phylogenetic analyses suggest that modern birds split off in the Cretaceous, but representatives from the ancestral lineage of birds are virtually unknown from the Mesozoic. The first paleognaths do not appear in the fossil record until about 60 million years ago, although they are actually thought to have existed much earlier. The situation is similar for the neognath birds. Key questions about their geographic distribution and ecology, and the actual divergence of modern birds, therefore remain unanswered. However, a Mesozoic fossil was recently discovered in Belgium that is undoubtedly a modern bird from the neognath group. The fossil fills an important phylogenetic gap in the early evolutionary history of the so-called crown birds. Some characteristic features of its skull indicate that the animal was closely related to the last common ancestor of chickens and geese and may be the last common ancestor of Galliformes and waterfowl—and hence, the first known representative of the Galloanserae. The find is from the latest Upper Cretaceous period and is about 66.8 to 66.7 million years old. Hence, it lived contemporaneously with *Tyrannosaurus* and *Triceratops*, about seven hundred thousand or eight hundred thousand years before the asteroid hit. My doctoral advisor, Daniel Field, studied it with scientists from Cambridge, Maastricht, and Greenwich. The fossil consists of an almost complete, three-dimensionally preserved skull and individual elements of the postcranial skeleton. This makes it the first modern bird from the Mesozoic with a well-preserved skull. It represents one of the few unequivocal pieces of evidence of crown birds from the Mesozoic and exhibits a unique combination of galliform and waterfowl features. Interestingly, it was also found in strata in which extinct tooth-bearing seabirds appear that do not yet belong to the crown group: relatives of *Ichthyornis*. The co-occurrence of this animal with *Ichthyornis* relatives may be evidence of their coexistence. It also challenges previous hypotheses that crown group birds originated in the southern hemisphere. The team named the small bird *Asteriornis maastrichtensis*. The genus name means "Asteria's bird," while the species name refers to where it was found and its geological age. Both names take into account other additional aspects of the fossil. Namely, Asteria was a Titan from Greek mythology who transformed herself into a quail to escape Zeus's advances. Quails are galliform birds and *Asteriornis* is the mother of all galliformes. Also, Asteria once plunged from the sky into the sea—just like the asteroid that fell into the sea off the coast of Chicxulub at the end of the Cretaceous period.

The importance of this find can hardly be overstated: first, because it provides us with new information about the habitat of the first modern birds; and second, because finds of the very first modern birds are extremely rare. In fact, there is only one other bird from the late Upper Cretaceous that undoubtedly belongs to the modern birds and of which more than a single isolated bone has been found. This bird is called *Vegavis* and is about 66.5 million years old, or about two hundred thousand to three hundred thousand years younger than *Asteriornis*. It was found on Vega Island, which is off the Antarctic Peninsula. This is where the name of the animal comes from. Some colleagues suggest that *Vegavis* is closely related to ducks and geese, while others believe it is still phylogenetically just outside the Galloanserae. To be certain of the answer to this question, we would need further skull material of the animal, through which we could recognize characteristic features. However, there is more evidence that it belonged to the neognaths than it being an ancestor to the group. In any case, *Asteriornis* and *Vegavis* have relatively small body sizes, and both finds come from sediments that indicate they were coastal dwellers. They lived in a similar ecosystem, which may provide an explanation for why modern birds were able to survive the asteroid impact. I return to this question in the last chapter of the book.

Beyond the Rainbow

Although we do not know exactly when the first feathers emerged, it is clear that some dinosaurs possessed them—long before birds could take to the skies with their help. So, feathers are not a unique feature of birds, and they have served many other purposes before bird flight. In modern birds, contour feathers form the majority of the visible plumage. The contour feathers covering the body are called the pennaceous feathers.

Those pennaceous feathers that are used for flying are fittingly called flight feathers. They are long, stiff, asymmetrically shaped, but symmetrically paired on the wings or the tail of the bird. The flight feathers on the wings are called remiges (singular remex), and those on the tail are called rectrices (singular rectrix).

In addition to the contour feathers, birds have down feathers that insulate the body. Their pennaceous feathers are necessary for flight, but also for communication, camouflage, and breeding. Which function the pennaceous feathers originally served in dinosaurs has long been unclear. We don't know exactly when dinosaurs first developed contour feathers; the oldest evidence comes from the Jurassic period. The remains of some *Archaeopteryx* specimens show that these animals also had contour feathers about 150 million years ago. Closely related to *Archaeopteryx* and even older is a fossil from Hebei Province in China, east of Beijing. It is about 161 million years old and has asymmetrical contour feathers. Whether or not *Archaeopteryx* was able to fly actively is still a matter of debate. However, it is undisputed that it was able to glide and flutter. This is shown by its almost modern-looking asymmetrical flight feathers and the clavicles which are fused into one wishbone. In an asymmetrical feather, the vanes on the left and right of the quill are of different widths. Such asymmetry is considered a mandatory requirement for powered flight. It gives the birds aerodynamic characteristics that are essential for flight. The strong wishbone also indicates that the animal's thoracic muscles were particularly robust and that the animal could perform powerful strokes with its wings. Whether this enabled *Archaeopteryx* to lift off the ground under its own power cannot be said with certainty. The backward-facing first toe of its foot could be evidence that it was an arboreal dweller that used this claw to climb or cling to branches. The trees may have served as a launching pad for when it was unable to get off the ground. However, the area of the Solnhofen Limestone in Bavaria, where these proto-birds (Urvögel) were found, are deposits of a lagoonal landscape that formed on the edge of the Jurassic Sea in a hot, dry climate, where there were few or perhaps no trees. *Archaeopteryx* would have had to climb high cliffs and swoop down from them to demonstrate its flying skills. In the eleventh specimen of *Archaeopteryx* discovered so far, some of my colleagues from Bavaria have discovered spectacular plumage that covered its entire body, even its legs, with long feathers, arranged in rows. In modern birds of prey, such long leg feathers do not actively contribute to flight but are helpful when landing. According to Christian Foth of the University of Fribourg, who studied this specimen with his team, the feathers may have not only served for flight, but mainly for display because of their smooth surface and light-refracting properties. Thus, contour feathers were superior to filamentous protofeathers in courtship. Protofeathers probably evolved in early dinosaurs. Initially, their main function was exclusively thermal insulation. Thus, they contributed to an increase in growth rate, which was accompanied by a faster metabolism at the beginning of dinosaurian evolution. This requires effective insulation of the body; otherwise, metabolic energy is lost as heat instead of being available for growth. This was especially crucial for small- to medium-sized early dinosaurs and their young, which had an unfavorable mass-to-surface ratio. While the long-necked dinosaurs and the tyrannosaurids grew larger and larger over the course of their evolution, the representatives of the avian lineage began a reduction in body size. This miniaturization with simultaneous rapid growth and high metabolism was only possible due to good body insulation.

Now it was the case that the dinosaurs had already developed colorful scales to attract mates or deter rivals. These scales must have been very colorful and arranged in species-specific patterns. The special arrangement of scales of different sizes suggests that these scales also had different colors. Therefore, dinosaurs may have had stripes like tigers or zebras, or spots like leopards, and at the same time exhibited such a beautiful play of colors as sand lizards. However, this blaze of color was limited by the first protofeathers—namely, the hairlike filaments entailed the loss of structural, color signaling capabilities. This is still the case today in birds, whose body coverings consist of flexible, filamentous feathers. Iridescent, variegated colors are only possible in tropical birds because they have contour feathers. Hair or protofeathers cannot produce such color effects. Displaying iridescent greens and blues or ultraviolet color reflections requires a precise arrangement of nanometer-scale light-scattering elements, which in modern bird feathers consist of keratin, melanosomes, and even air. It is through the tiny but highly ordered rami and barbules in vanes, interlocked by tiny hooks, that a bird produces the iridescence and rich play of colors of its contour feathers. These colors are essential for communication and sexual selection. Thus, to maintain the rich blaze of color of their scales, tiny dinosaurs could not simply rely on protofeathers, but had to evolve more complex feathers that included keratin, melanosomes, and air in their structure. At the same time, the colorfulness of feathers is only as good as the visual ability of their counterparts allows—which raises the question of the visual color spectrum of dinosaurs. We know that virtually all of today's reptiles can perceive colors in a much more sophisticated way than humans and other mammals. Many reptile species can even see ultraviolet colors. We are not able to do that. If we infer phylogenetically, this suggests that dinosaurs were also equipped with the highly differentiated color vision of birds, known as tetrachromacy.

Cone cells in the retina of the eye are responsible for the perception of colors. Cone cells are neurons that are needed as specialized sensory cells for seeing in daylight. Mammals have three different cone cells and are therefore called trichromats. Reptiles have four different cones, which is why they are called tetrachromats. Trichromats can recognize about a million different colors, and tetrachromats up to a hundred times more. It is believed that they can distinguish about 100 million colors. Three of their cone cells perceive the colors blue, green, and red, and via a fourth, short-wave receptive neuron they can recognize the ultraviolet color spectrum up to shades of turquoise.

Thus, pennaceous feathers may have solved another very complex evolutionary problem in addition to enabling powered flight. Maniraptors, for example, independently evolved feathers in different branches that were clearly not for flight but for good insulation, to facilitate increased metabolic rate, and miniaturization without losing the structural color signaling of their scales. This would not have happened with a complete covering of protofeathers or hairs. This is because fur, unlike feathers, has a limited color spectrum, because without a coherent surface it does not enable directional light scattering and compromises structural color representation. The color pigments in hair therefore produce only reddish-brown to black tones, with lighter colors achieved by the absence of pigments.

In mammals, too, there was a trend toward reducing body size and increased thermal insulation through thick fur. However, because mammals were nocturnal in the Mesozoic era, sophisticated color perception did not play a role. At night, all cats are gray. Tetrachromacy is probably an original feature of terrestrial vertebrates, and in the ancestors of modern mammals differentiated color vision was lost because it had no meaning at night and played no role in their survival.

In dinosaurs, however, flat or planar structures gradually evolved from hairlike feathers. Planar feathers initially obscured the filamentous protofeathers and replaced them over the course of evolution, eventually leading to the development of smooth contour feathers. And while we do not know exactly when the first protofeathers appeared, it seems at least clear that the first planar contour feathers occurred in the Maniraptora group. In them, the rami and barbules of the contour feathers were in one plane. The hooks and barbs, with which the barbules were held together, could, just like in a zipper, be opened and closed as often as desired. This made the feathers extremely durable despite their lightweight construction. In addition, the fact that the rami and barbules were positioned closely next to each other on a single plane created a particularly smooth, flat, and uniform surface. This was particularly important because the precise arrangement of the light-scattering pigments allowed optimum light penetration, which was a prerequisite for the iridescence of the feathers. The birds achieve the best color play only with these flattened filaments. The smooth, foil-like surfaces provide optimal physical conditions for the iridescent rainbow colors. It is surely no coincidence that the maniraptors, of all dinosaurs, who had very mobile wrists, also had large feathers on their arms and hands that allowed them to maximize surface area and create structural color signaling on a macroscopic level. Large, smooth feather surfaces acted as a canvas for detailed ornamentation and patterns.

Ostriches, which have frizzy plumage, as opposed to a well-structured planar plumage, have only brown, black, or white integuments (just like mammals), and cannot produce the complex variety of colors seen in contour feathers. If we remember that paleognaths are the only modern birds that distinguish colors poorly, this also makes sense. A special case among paleognaths is the helmeted cassowary, which can probably see more colors than its relatives. The necks of cassowaries shine a bright blue that comes not from the feathers, but from the strong skin color on the neck. The fact that paleognaths are the only birds today that have filamentous feathers should definitely be given greater consideration in future phylogenetic analyses. Might it be possible that they split off from coelurosaur ancestors even before the Maniraptora evolved? If they underwent convergent evolution, would we recognize that they merely resembled the Maniraptora and other present-day birds? Would we recognize that they do not have the same maniraptoran ancestor as the neognath birds—and that even the paleognaths are descended from different coelurosaurs, when their closest relative outside of birds is the crocodile?

In this context, the compsognathids, a group of very small carnivorous coelurosaurs, have caused a stir in research circles. In a compsognathid from China, a lush, fur-like plumage was discovered. Fossil melanin was found in the compsognathid *Si-*

nosauropteryx, whose name means "Chinese reptile wing." The melanin allowed its plumage color to be determined. Lacking planar feathers, this animal did not have a wide range of iridescent colors. Instead, during its lifetime, its fibrous feathers were brown to reddish brown in some places and white in other places. What made this find special, however, was not the colors alone, but a countershading, as we have already encountered in the armored *Borealopelta*. *Sinosauropteryx* comes from the Early Cretaceous strata of the Jehol ecosystem in northeastern China, which was previously thought to have consisted of dense, closed forest. Characteristic color patterns in the plumage of the dinosaurs that lived there may confirm this, because light conditions in the forest were different from those in open terrain. However, the coloration of *Sinosauropteryx* does not indicate a forest dweller, but rather an unprotected, wide-open habitat. To be sure that the colors of *Sinosauropteryx* were indeed fossil melanin, two animals of this species were examined, and intriguingly, nearly identical color patterns were found in both. The bridge of the nose, the lower jaw, the throat, and the chest were white, but the area around the eyes was dark. Thus, *Sinosauropteryx* looked almost like a raccoon. However, such "bandit's masks" are not only typical for raccoons, Zorro, or the Beagle Boys (from Donald Duck cartoons); they are a common pattern in many birds, where they serve as camouflage. The tail of *Sinosauropteryx*, on the other hand, was very reminiscent of the curly tail of a lemur, with alternating white and reddish-brown bands. In any case, color reconstruction helps to gain a better understanding of habitat, and we can tell more about ecology from contour shading than from skeletal features alone. The fact that small predatory dinosaurs also lived in open habitats in the Jehol bioprovince contributes to a clearer picture of the environment and fauna, and therein may lie the key to the great biodiversity in the Jehol Basin. The extraordinary species richness of this area can be more easily explained if we consider that this region has evolved over a period of more than 10 million years, and that the vegetation, as well as the landscape, has changed drastically several times during this period. It is therefore not a contradiction when we find both forest and steppe dwellers in the Jehol bioprovince, where the most spectacular feathered dinosaurs of China come from. The biodiversity is indeed remarkable. Here lived not only enantiornithines, but also even more basal, small dinosaurs, such as microraptors, which wore feathers but were not yet birds even in the broader sense. They were descended from the sickle-clawed raptors. Another small feathered dinosaur is called *Confuciusornis*. They are also not yet a modern bird, but are at least already on the evolutionary branch that led to birds. Some of them have extravagant decorative feathers on their tails and at least outwardly remind us a little of birds of paradise. There are hundreds of fossils of these animals, and Chinese researchers were able to determine a sexual dimorphism in them, where one sex type had particularly long and extravagant tail feathers and the other did not.

What *Microraptor*, *Confuciusornis*, the enantiornithines, and the "almost-birds" of Jehol have in common are their contour feathers. Unlike *Sinosauropteryx*, they all already had smooth, planar

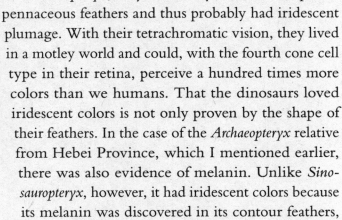

pennaceous feathers and thus probably had iridescent plumage. With their tetrachromatic vision, they lived in a motley world and could, with the fourth cone cell type in their retina, perceive a hundred times more colors than we humans. That the dinosaurs loved iridescent colors is not only proven by the shape of their feathers. In the case of the *Archaeopteryx* relative from Hebei Province, which I mentioned earlier, there was also evidence of melanin. Unlike *Sinosauropteryx*, however, it had iridescent colors because its melanin was discovered in its contour feathers, which is why the animal became known as the "rainbow dinosaur." The name of this newly discovered dinosaur species is

Caihong juji, which is Mandarin and means "the rainbow with the big crest." This term refers to the animal's large, broad, and iridescent feathers and the bony crest it wore above its eyes.

The platelet-shaped melanosomes of *Sinosauropteryx* were discovered for the first time under a scanning electron microscope. They were still intact and could be compared with those of modern birds. The feathers of *Caihong* bear a strong resemblance to those of hummingbirds. Their melanosomes are actually black, but light is refracted by the special platelet shape to create a shimmering effect. Modern birds use their colorful plumage to attract mates. Accordingly, *Caihong*'s rainbow feathers could be a prehistoric version of a peacock's iridescent tail. *Caihong* is the oldest known example of platelet-shaped melanosomes typically found in bright, iridescent feathers. It is also the oldest animal known to date with asymmetrical feathers.

CHAPTER 13

The End of the Dinosaurs

The Day it Rained Glass

On an otherwise uneventful day in spring 66 million years ago, a huge boulder from space, the size of New York, crashed into Earth. As fascinating as the dinosaurs, the flying reptiles, and the marine reptiles were, they were nevertheless wiped out by this asteroid impact at the end of the Cretaceous period. They did not manage to escape the cosmic catastrophe. While the birds' survival also ensured the continuation of asymmetrical feathers, other innovations of nature were lost forever. Today, there is no animal that can fly underwater like a plesiosaur with four paddle-shaped fins. There are no flying animals that have wingspans of more than thirty or so feet, or long-necked giants that weigh more than fifty tons. Crocodiles cannot gallop today, and there are no carnivores that weigh fifty times more than modern lions. Some ecological strategies have also been lost, such as the splitting of an animal species into different ecological niches, as we have seen in long-necked dinosaurs and tyrannosaurs. This strategy, which was incredibly successful over millions of years, ended up sealing their fate.

Traveling at a speed of fourty-four thousand miles per hour

or more, and with a diameter of more than six miles, the asteroid hit so hard that it even melted the Earth's crust, locally. The
extraterrestrial object crashed into Earth off the coast of the Yu-
catán Peninsula in Mexico. The small village Chicxulub is the
closest human settlement to the impact site, which is why the
asteroid is now called "the Chicxulub impactor." The impact
was followed by tremendous shock waves that formed tsunamis
in the Gulf of Mexico that were more than a mile high. In Big
Bend National Park in Texas, about a thousand miles from this
site, we find sediments that were formed immediately after the
impact and are called tsunamites.

> **Tsunamites are sedimentary rocks composed of the
> sediments formed during a tsunami.**

The impact of the asteroid not only caused earthquakes and
tsunamis, but also ejected huge amounts of debris, molten rock,
and dust from the crater. The force of the impact was so violent
that some pieces of rock may have even been hurled into space.
At the Tanis Fossil Site in North Dakota, about two thousand
miles from Chicxulub, researchers have found tektites associated with the asteroid.

> **Tektites are glass objects formed when surrounding
> rock melts during an asteroid or meteorite impact
> and is hurled away from the impact site. The mostly
> teardrop-shaped objects can grow to several inches
> in size and are often composed of high-pressure mod-
> ifications of quartz.**

In the Tanis sediments, which are part of the Hell Creek Formation, fish skeletons were found that had been hurled ashore
and perforated by small tektites. In the fossilized remains of
some of these fish, the small projectiles were still stuck in their

gills. The fish were an important indication of the seasonal confinement of the impact event. Indeed, the earthquake deposits show an annual cyclicity in the last years of the Cretaceous and prove that the impact occurred during the northern hemisphere's springtime.

I remember the early phase of the research project surrounding the Tanis fossil site well—I was not involved myself, but was lucky enough to meet Melanie During, the lead author of the paper, in Grenoble in 2018. I was scanning Mesozoic and recent fish with our Oxford team while Melanie was there segmenting fish fossils and the small, spherical tektites for her project. I had seen her at a few meetings, but I had never talked to her before. In Grenoble, she took the time to explain her work to me, which she published in *Nature* in 2022. I found the topic exciting at the time, not realizing that her research would later receive so much attention. As it turned out, she could detect seasonal changes in the bones of the fish, just as we have seen in the bones of *Thrinaxodon*. Just like *Thrinaxodon*, there were signs of aestivation and hibernation in the fish. In addition, their reproduction apparently followed annual cycles, and the food supply also varied depending on the season, which of course makes sense, since otherwise aestivation and hibernation would not have been necessary.

It is amazing that the Tanis site provides an accurate chronology of the immediate events of the Chicxulub impact. The strata there were formed within about an hour of the impact and represent—far inland—a mixture of marine and terrestrial deposits. Although the nature of their assemblage is reminiscent of tsunamites, these layers were likely formed by the shock waves of a massive earthquake. Researchers estimate the earthquake that could lead to such strata had a magnitude between 10 and 10.6 on the moment magnitude scale (M_w). A 10 M_w earthquake already releases the energy of 1.2 million Hiroshima bombs. The most violent earthquake ever measured in our time occurred in Chile in 1960 with 9.5 M_w. At first, this may sound similarly devastating as the primeval asteroid impact; however, an earthquake with a magnitude of 10.5 M_w is about thirty-six times more powerful than one with 9.5 M_w. The moment magnitude scale ends at the value of 10.6 M_w, because physicists assume that the Earth's crust would break apart completely. So, if the asteroid had been just a little bit bigger and crashed into Earth a little bit faster, it might have destroyed the whole planet.

Evidence of the dust cloud caused by this impact, which completely enveloped the Earth for a short time, are wafer-thin layers of sediment known as the iridium anomaly. In 1980, this anomaly, detectable around the world, was scientifically described by Luis and Walter Alvarez, Frank Asaro, and Helen Michel. The team was able to detect an elevated concentration of iridium in a boundary layer between Cretaceous and Paleocene rock deposits in Italy and Denmark. Such layers have since been detected in New Mexico, New Jersey, and other regions. In 2022, an article was published that reported on rock layers at the Cretaceous–Tertiary boundary from Baja California, Mexico. They consist of terrestrial and shallow marine sediments that were rapidly rebedded onto sediments on the continental slope as a result of the impact. Radiometric dating indicated that the strata are about 66.12 million years old, with a margin of error of 650,000 years. This corresponds to the age of the Cretaceous–Paleogene boundary. They contain corals, marine gastropods, and bivalves, as well as tuffs, quartz, and charred logs that originated on land. The quartz grains are shocked quartz, formed only when earthquake waves pass through them. This unusual mixture of disparate materials is interpreted as heterogeneous landslide deposits generated by an earthquake as a direct result of the Chicxulub impact and a mega-tsunami. The charred logs in the sediments probably formed in a very short period of time. Immediately after the impact, a huge fire wave with a temperature of more than 1,832°F must have swept over the land at this location, which was extinguished by the tsunami only a few minutes later. As the water flowed back, the floods then swept away trees, tuffs, and coastal sediments, washing them into deeper water. Directly above, we now find mudstones that also bear a distinct iridium signature. While the Tanis sediments were formed within an hour of the impact, these sediments were deposited within the first ten minutes after the impact.

Since the element iridium hardly occurs in the Earth's crust, but is found in high concentration in meteorites, an iridium-rich layer can be an unmistakable sign of an extraterrestrial impact. Some astronomers believe that the origin of this asteroid was in the farthest reaches or just outside our solar system in the Oort Cloud. Recent evidence suggests that the asteroid may have broken apart shortly before the impact, and a second part of the celestial body crashed to Earth off the west coast of Africa.

Like a Phoenix From the Ashes: How Birds Survived the Catastrophe

When the asteroid hit the Earth 66 million years ago, all dinosaurs except modern birds became extinct, as well as the pterosaurs, all marine reptiles, and the ammonites. The asteroid impact also had a massive impact on plant life. Several factors caused global deforestation. The shock waves that went around the world immediately after the impact certainly caused significant forest loss, and subsequent forest fires devastated large areas of vegetation. In addition to the immediate effects of the impact, there were long-lasting factors that damaged the forest. Even in the months that followed, ash, smoke, and soot in the atmosphere caused reduced solar radiation, which massively impaired plant photosynthesis while causing global cooling. The entire world was in a permanent dim state. In the twilight, forests continued to die. Even as the dust slowly settled, low sunlight and cooler temperatures delayed recovery in massively damaged forests. Why forest dieback and ecological collapse were less severe for birds than in their closest relatives is still the subject of extensive debate. And as simple as the question of why the birds survived may sound, the answer seems very complex. There are two ways to ask this question. We can ask once why all other dinosaurs could not recover from the asteroid impact or in what way the birds differed from the other dinosaurs that allowed them to survive.

To answer the question of survival, we must first know when exactly modern birds evolved. The long ghost lineage among ostrich relatives makes this surprisingly difficult. The first representatives of the group may have evolved more than 100 million years ago, but we find fossil evidence of the group only shortly before the Cretaceous–Paleogene boundary. The Galliformes and waterfowl originated only a few hundred thousand years before the Chicxulub impact, and all other birds probably thereafter. But there is hardly any fossil evidence of the early representatives of modern birds. This is partly because birds are usually very small animals with delicate bones and are often not preserved in the fossil record. Moreover, they only play a very minor role in ecosystems at the end of the Cretaceous, yet.

The first reliable evidence of a true Neoaves is dated to 62.5 million years ago—that's about 3.5 million years after the asteroid impact. The fossil came from the state of New Mexico and was named *Tsidiiyazhi abini*, a Navajo term meaning "the little morning bird." This animal is believed to be related to today's mousebirds, which are found exclusively in Africa, south of the Sahara. This complicates our question, since mousebirds are closely related to passerines, and genetic studies suggest that mousebirds and passerines must actually have evolved later than many other Neoaves. Most phylogenies suggest that flamingos and pigeons were the first "new birds." Then came hawks, hummingbirds, cranes, gulls, penguins, and many others. The core land birds, which include the mousebirds and passerines, were probably the last group of Neoaves. If we look more closely at the earliest Cretaceous fossil record, we find that Galliformes and waterfowl lived in very different habitats than the first Neoaves *Tsidiiyazhi*. *Asteriornis* and *Vegavis* were more coastal dwellers. And the first known waterfowl, just like *Vegavis*, comes from Antarctica. The animal is called *Conflicto antarcticus* and was deposited in strata of the López de Bertodano Formation, dating from the Late Cretaceous to the earliest Paleocene. The layers in which it was found indicate that the bird lived shortly after the asteroid impact. But why were these early modern shorebirds or waterbirds able to survive while the ichthyornithids and enantiornithines were not? One of the reasons surely lies in the fact that a number of their characteristics proved to be selectively advantageous after the mass extinction. These included the lower energy requirements of the birds. Because of their small body size, they also had lower energy requirements than most other dinosaurs, while showing a certain adaptability and flexibility in their diet. With the collapse of many ecosystems, considerable food resources disappeared, and the earliest birds had to rapidly change their dietary habits. Perhaps an advanced digestive system helped them do this, allowing them to switch their diet to insects and seeds. For the small birds at that time, just a few insects or a few seeds a day were probably enough to survive. Larger animals, on the other hand, did not find enough food. The ichthyornithids, which are most closely related to today's birds, were much larger and probably predominantly fish eaters. But the fish faunas were severely damaged by the impact, so it became increasingly difficult for these animals to find enough food. At the same time, their energy requirements were higher because they were larger and heavier, and their specialized fish diet did not allow them to switch quickly enough to seeds and insects. In the case of enantiornithines, their continued existence was probably affected by the destruction of forests. They could not adequately protect and feed their chicks, which were predominantly precocial.

The other flightless dinosaurs could not survive because of their high energy demands. However, were the conditions really equally bad everywhere in the world? Were there not perhaps still isolated small island populations which had been spared by the catastrophe? Did these islands maybe also offer somewhat better conditions after the impact? We have already indirectly addressed the answer to these questions several times. Dinosaurs were very well adapted to their environment, and in a stable ecosystem they could excellently assert themselves against competitors and displace them. But it was this successful strategy that caused their demise in an ecological crisis. In mammals and modern birds, offspring and adults occupy the same ecological niche, so survival was possible for the young only if their parents found enough food. In dinosaurs, and especially in large predatory dinosaurs and long-necked dinosaurs, the different generations behaved, trophically, like different species. As a reminder, the different diets of the young and old led to a reduction in niche diversity because hatchlings, juvenile large predators, and adults functioned as distinct ecological morphospecies. The same was true of long-necks, whose young fed on horsetails, while adult sauropods ate leaves from tall trees. If the young did not find enough food, they did not grow and starved to death before sexual maturity, and if the adults did not find food, they died before they could lay eggs.

There were selective advantages for smaller mammals because, like birds, they often fed on insects. Larger mammalian species could not survive either. Because non–avian dinosaurs occupied all ecological niches during their lifetime, none of them could survive.

But even for small mammals and birds, resources were scarce. When their hunger eventually became stronger than fear, it drove them to selectively scavenge dinosaur clutches, which only accelerated their extinction.

EPILOGUE

This book is intended to give an overview of the incredible discoveries in paleontology and to show the fascinating findings that my colleagues and I have made in recent years. Research is active and we are able to use ever new methods to unravel mysteries that previously seemed unsolvable. However, the book contains only a tiny sample of what is currently being researched, and if it finds appeal, I will be happy to tell more in another book. For the stories I have told, information has flowed together from different corners of the globe. Whereas fifty years ago research was dominated by North American and European paleontologists, today the most exciting discoveries often come from South America, Asia, Africa, Australia, and Antarctica. We understand dinosaurs better than ever before, and yet there have never been so many unanswered questions. With every door we push open, we face new puzzles to unravel. In the process, understanding the dinosaurs also helps us better understand our own destiny and perhaps steer it in the right direction for the benefit of our children.

Three of the largest mass extinction events on Earth are related to the emergence, evolutionary success, and demise of the dinosaurs. Due to a constantly increasing concentration of greenhouse gases, galloping effects occurred, first creeping and then accelerating. Ecological collapse was then always followed by massive species extinction, triggering a trophic cascade. Birds survived the last catastrophe because they evolved in ecological niches that were least affected by changing environmental conditions. Today, humans are destroying bird habitats and food sources and threatening biodiversity on such a scale that we may be responsible for the next great mass extinction. If we are not careful, we will complete what the asteroid failed to do and ensure the extinction of the last dinosaurs on Earth. But that need not be the case. We know and understand what is happening around us. Now it is up to us to take the right actions and listen to those who understand science and offer us solutions to these challenges.

★ ★ ★ ★ ★

GLOSSARY

A

aestivation: a form of summer hibernation in which the metabolism is lowered and all activity is suspended or shut down altogether as an adaptation to adverse environmental conditions. It can occur during periods of heat or drought, and seasonally in arid climates. Aestivating behavior still exists today in some moth and snail species.

Albian: the last chronostratigraphic stage of the Lower Cretaceous (112.9 to 100.5 million years before present).

Alcmonavis: a prehistoric bird from sediments of the Upper Jurassic of the Altmühl valley in Bavaria (about 150 million years old).

Allosaurus: a genus of theropod dinosaurs that lived during the Upper Jurassic of North America and Europe and was one of the largest predators of its time.

altricial birds: birds that live in their parents' nest for an extended period of time after hatching and require brood care.

ammonoidea: a group of extinct cephalopods that lived 407.6 to 66 million years ago.

Amniotes: a large clade that includes all terrestrial vertebrates, with the exception of amphibians, that is characterized by the ability to reproduce in locations outside of the water.

anagenesis: the evolution of a new species that replaces an old species.

Anisian: the first chronostratigraphic stage of the Middle Triassic (247.2 to 242 million years before present).

anoxic (water): fresh or salt water that is entirely depleted of oxygen.

Archosauria: a clade of tetrapods that includes crocodilians, birds, dinosaurs, and pterosaurs and probably originated in the late Upper Permian.

Asteriornis: a genus of birds from the Late Cretaceous of Belgium (about 66.8 to 66.7 million years before present), last common ancestor of all chickens and geese and thus the oldest known neognath bird.

Aves: the scientific name for the class of birds.

Avialae: the group of theropods more closely related to birds than to deinonychosaurs.

B

Baryonyx: a dinosaur from the family Spinosauridae, from the Lower Cretaceous of England (about 130 to 125 million years before present).

biostratigraphy: the correlation and relative chronological classification of rock strata based on fossils.

bioturbation: the burrowing of soils and sediments by burrowing organisms.

bone beds: an accumulation of fossils in certain rock layers.

Bone Wars: a term used in the American press and popular science literature to describe the dispute between two American paleontologists, Edward Drinker Cope and Othniel Charles Marsh, that lasted from 1877 to 1892.

Borealopelta: a genus of armored dinosaurs from the Lower Cretaceous of Canada (112 to 110 million years before present), whose skeleton is the most complete ever found of this group.

Brachiosaurus: a genus of long-necked dinosaur from the Upper Jurassic of North America and Africa (154 to 145 million years before present) found in the Morrison and Tendaguru formations.

Brontosaurus: a genus of long-necked dinosaur from the Upper Jurassic of North America found in the Morrison Formation.

Broomistega: a genus of amphibians from the Early Triassic of South Africa.

C

Caihong: a genus within the group Avialae.

Camarasaurus: a genus of long-necked dinosaur from the Upper Jurassic of North America found in the Morrison Formation.

Ceratopsia: the group that comprises all horned dinosaurs.

Chicxulub: a municipality in Mexico, near the impact site of the asteroid that hit Earth 66 million years ago.

Coelurosauria: a clade within Theropoda.

cladogenesis: a phenomenon whereby an original species splits into two new species.

computed tomography: an imaging procedure in radiology that can be used to x-ray bones and fossils.

concretion: an often round or spherical mineral composite. It consists of a hard, fine-grained sediment that has been baked by pore water and grown outward from a crystallization center over time.

confidence interval: an interval in statistics used to indicate the accuracy of the location of a parameter. The confidence interval indicates the range that includes with a certain probability the parameter of a distribution of a random variable. A confidence level of 95 percent is often specified by researchers to measure the informative value of data. When a confidence interval is calculated, it then includes the true value with a probability of 95 percent.

Conflicto: the earliest genus of geese.

Conodonta: a clade of basal vertebrates.

convergent evolution: when completely unrelated animals evolve to share a similar body shape, even though they did not descend from the same ancestor.

Cope, Edward Drinker (1856–1897): an American vertebrate

paleontologist, zoologist, comparative anatomist, herpetologist, and ichthyologist.

coprolites: trace fossils consisting of feces, which are usually phosphatic.

Cretaceous: the youngest period of the Mesozoic (145 to 66 million years before present).

Cymbospondylus: a genus of marine reptiles from the ichthyosaur clade.

cynodontia: a group of early mammal-like therapsids.

D

Daonella: a genus of free-swimming marine bivalves from the Triassic period.

dicynodontia: a group of herbivorous synapsids.

dinosauria: the largest group of terrestrial vertebrates of the Mesozoic era.

Dinosauromorpha: a major group within the archosaurs that includes the dinosaurs and their closest relatives.

Diplodocus: a genus from the group of long-necked dinosaurs from the Upper Jurassic found in the Morrison Formation.

E

ecology: the study of living things' relationships to each other and their environment.

Edmontosaurus: a genus of duckbill dinosaurs.

Enantiornithes: a sister group of modern birds, that went extinct 66 million years ago.

Eocene: a chronostratigraphic series of the Paleogene (56 to 33.9 million years before present).

Europasaurus: an island-dwarfed sauropod from northern Germany.

F

fossil deposit: a site that is particularly rich in fossils.

fossil record: a documentation of all occurrences of fossils that places them in a scientific, stratigraphic context.

fossils: the remains of living things and evidence of their activity that are at least ten thousand years old and can attest to past life in Earth's history.

G

Galloanserae: a major group of modern birds, including all fowl and geese.

Germanic Basin: a large sedimentary area that extended from England to eastern Poland during the Permian and the Triassic and was at times covered by a shallow sea.

ghost lineage: a gap in a species lineage from a hypothesized ancestor in a species that has left no fossil evidence, but can still be inferred to have existed because of its descendants.

Gondwana: a southern supercontinent that consisted of present-day Africa, South America, Antarctica, India, and Australia.

H

Helveticosauridae: a family of basal archosauromorph marine reptiles from the Middle Triassic of Switzerland.

Heyuannia: a genus from the oviraptorian group from China, of which innumerable clutches are known.

Holtz, Thomas Richard, Jr. (*1965): an American vertebrate paleontologist with the University of Maryland.

I

Ichthyosauria: a group of extinct secondary marine reptiles that were fully adapted to life in the sea.

Iguanodon: a genus in the group of ornithopod dinosaurs.

island dwarfing: a phenomenon in evolutionary biology whereby the size of animals in an isolated ecosystem decreases within a few generations. It occurs especially on islands where these animals do not have to fear predators. Other remote habitats such as caves, cut-off valleys, or inaccessible mountainous regions can also cause this phenomenon.

iridium anomaly: a globally detectable elevated concentration of the element iridium in sedimentary rocks deposited 66 million years ago, at the Cretaceous–Paleogene boundary, which is evidence of an asteroid impact.

J

Janensch, Werner Ernst Martin (1878–1969): a German vertebrate paleontologist, geologist, explorer, and one of the foremost dinosaur specialists of his time, best known for his African expedition.

Jurassic: the middle period of the Mesozoic (201.3 to 145 million years before present).

K

Kimmeridgian: the middle stage of the Upper Jurassic (157.3 to 152.1 million years before present).

L

Leidy, Joseph (1823–1891): an American paleontologist and professor of anatomy with the University of Pennsylvania.

M

Maastrichtian: the latest stage of the Upper Cretaceous (72 to 66 million years before present).

Maniraptora: a derived group of the Coelurosauria.

Maraapunisaurus (formerly *Amphicoelias*): a genus of long-necked dinosaurs from the Upper Jurassic of North America found in the Morrison Formation. The animal is considered a hot contender for the record of the largest dinosaur ever found.

Marsh, Othniel Charles (1831–1899): an American vertebrate paleontologist and the first professor of paleontology at Yale University.

Megalosauridae: a family within the group Megalosauroidea.

Megalosauroidea: a superfamily within the group of theropods.

Megalosaurus: a genus of theropods from the Middle Jurassic of England (about 166 million years before present) that, in 1824, became the very first scientifically described dinosaur.

melanosomes: the specialized functional systems in pigment cells that produce the pigment melanin.

Mellisuga: an extant genus of hummingbirds and the smallest bird in the world.

Mesozoic: the second-to-last era of Earth's geological history, also known as "Earth's Middle Ages" (251.9 to 66 million years before present).

Mixosauria: a group of small ichthyosaurs that lived during the Middle Triassic.

N

Neoaves: the group of all modern birds except the Paleognathae and the Galloanserae.

Neognathae: the group of all modern birds except the Paleognathae.

Neovenator: a dinosaur closely related to *Allosaurus* and dated to the Lower Cretaceous (130.7 to 126.3 million years before present).

Norian: the middle chronostratigraphic stage of the Upper Triassic (228 to 208.5 million years before present).

Notatesseraeraptor: a genus of small theropods from the Triassic closely related to *Dilophosaurus*. The first carnivorous dinosaur described from Switzerland.

O

Odontocetes: the scientific name for toothed whales.

Oligocene: the uppermost series of the Paleogene (33.9 to 23 million years before present).

Omphalosaurus: an unusual ichthyosaur from the Triassic with crushing dentition.

Ornithischia: the scientific name for the bird-hipped dinosaurs.

Ornithopoda: a clade of ornithischian dinosaurs that started out as small, bipedal running grazers, and evolved into large duck-bill dinosaurs.

Osborn, Henry Fairfield, Sr. (1857–1935): an American vertebrate paleontologist, geologist, and president of the American Museum of Natural History.

osteoderms: dermal bones that usually form a carapace and serve to protect an animal.

Ostrom, John Harold (1928–2005): an American vertebrate paleontologist and discoverer of the predatory dinosaur *Deinonychus*.

Ostromia: a genus of small animals from the group of Maniraptora.

Owen, Sir Richard (1804–1892): an English paleontologist, biologist, comparative anatomist, and head of the British Natural History Museum in London, who coined the term "Dinosauria."

Oxfordian: the oldest stage of the Upper Jurassic (163.5 to 157.3 million years before present).

P

pachyostosis: a secondary adaptation of the descendants of terrestrial vertebrates, acquired through evolution, to life in water, in order to compensate for the static buoyancy of the body and to be able to remain underwater without effort. Of animals still living today, only manatees show pachyostosis, which is particularly noticeable in their broad, thick, and dense rib bones.

Pachypes: an ichnospecies of four-legged, five-toed herbivores from the Permian.

Paleognathae: the group of all modern birds except the Neognathae and the Galloanserae.

paleobond: an adhesive used to glue rocks, fossils, and petrified wood.

Paleogene: the oldest geochronologic period of the Cenozoic Era (66 to 23 million years before present).

Pangaea: a primitive continent that united all landmasses of the Earth in itself in the Permian.

Permian: the youngest period of the Paleozoic (298.9 to 251.9 million years before present).

paedomorphosis: the retention of juvenile characteristics at later life stages. It may be achieved by accelerating sexual maturity relative to the rest of the development or by delaying physical development relative to reproductive capacity.

Phalarodon: an early genus of ichthyosaurs discovered in Nevada and belonging to the mixosaur group.

phylogeny: within biological systematics, phylogeny describes the phylogenetic development of all living organisms and the relationship of the groups to each other.

Plateosaurus: a basal genus of Sauropodomorpha from the Upper Triassic.

pneumatization: the formation of air–filled cavities in originally dense bone tissue. Pneumatization can occur during the course of evolution as well as the course of individual development. Such bones are also referred to as "air–filled bones."

Potamornis: an extinct bird from the group of Hesperornithes, from the latest Cretaceous.

precocial birds: newly hatched young that are sufficiently developed to leave the nest and immediately follow their parents.

Proceratosauridae: a family of the group Tyrannosauroidea and sister group to the Tyrannosauridae.

R

Rauhut, Oliver (*1969): a German vertebrate paleontologist at the University of Munich.

S

Sander, Paul Martin (*1960): a German vertebrate paleontologist at the University of Bonn.

Saurischia: the scientific name for the group of lizard-hipped dinosaurs.

Sauropoda: the group of long-necked dinosaurs.

Sauropodomorpha: one of the two groups within the saurischian dinosaurs.

scanning electron microscope: a microscope that passes an electron beam over an object in a specific pattern. The interactions of the electrons with the object are used to generate an image.

serration: a sawtooth-like cutting edge on teeth.

sexual dimorphism: the externally visible differences between sexually mature male and female individuals of a species that do not relate to sexual organs.

Shonisaurus: a genus of ichthyosaurs discovered in Nevada and among the largest ichthyosaurs ever found.

Silesaurus: a genus of Dinosauromorpha from Poland and one of the closest relatives of the dinosaurs.

sclerotic ring: a bony, ring-shaped structure around the eyes of birds, dinosaurs, pterosaurs, ichthyosaurs, and iguanas.

Spinosauridae: a family from the group Spinosauroidea.

Spinosaurus: a genus of the Spinosauridae family with dorsal sails and one of the largest carnivorous dinosaurs, which has been found in Morocco and Egypt.

Stegosaurus: a genus of the ornithischian dinosaur group with bony plates on the back and a spiny tail found in the Morrison Formation.

Sternberg, Charles Hazelius (1850–1943): an American fossil collector.

synapomorphy: a derived characteristic that groups species together and distinguishes them from other species.

Synapsida: a large group of the amniotes.

synchrotron: a special type of particle accelerator.

T

Tanystropheus: a genus of basal Middle and Upper Triassic archosauromorphs whose greatly elongated neck was at least as long as their trunk and tail combined.

tsunamites: sedimentary rocks composed of the sediments formed during a tsunami.

tektites: glass objects that are formed when surrounding rock melts during an asteroid or meteorite impact and is hurled away from the impact site. The mostly teardrop-shaped objects are several centimeters in length and consist of modifications of quartz.

Temnospondyls: a group of terrestrial vertebrates that may be the ancestors of modern amphibians.

tetrachromacy: a type of color vision with four different cone cells. Reptiles are tetrachromats. It is believed that tetrachromats can see up to 100 million colors. Tetrachromacy is an original characteristic of terrestrial vertebrates.

Tethys: a Mesozoic ocean, located between the ancient continents of Gondwana and Laurasia, that completely closed during the Cenozoic, about 50 million years ago.

Tetrapoda: the large group of all terrestrial and secondarily aquatic vertebrates.

Thalattoarchon: a large predatory genus of the ichthyosaur group.

Therapsida: a large group of terrestrial vertebrates that includes mammals.

Theropoda: a group of saurischian dinosaurs that includes all carnivorous dinosaurs.

Thrinaxodon: a genus from the group of therapsids.

Tithonian: the uppermost stage of the Upper Jurassic (152.1 to 145 million years before present).

Torvosaurus: a genus of the family Megalosauridae.

Triassic: the oldest period of the Mesozoic (251.9 to 201.3 million years before present).

Triceratops: a genus from the group of horned dinosaurs.

trophic level: a particular level within the food chain in an ecosystem.

Tuebingosaurus: a genus of a sauropodomorph dinosaur from the Late Triassic of Germany.

tuffs: volcanic igneous rocks.

Tyrannosauroidea: a superfamily from the group Theropoda.

Tyrannosauridae: a family from the group Tyrannosauroidea.

Tyrannosaurus: a genus of the group Tyrannosauridae.

Thyreophora: a group of ornithischian dinosaurs that includes all stegosaurs and ankylosaurs.

Y

Yutyrannus: a genus of the group Tyrannosauroidea.

Z

zircon: a silicate mineral with the chemical formula $Zr[SiO]_4$.

REFERENCES

Chapter 1: New Life in the Sea

Wintrich, T. et al. An enigmatic marine reptile—the actual first record of *Omphalosaurus* in the Muschelkalk of the Germanic basin. Journal of Vertebrate Paleontology (2017).

Fröbisch, N. B. et al. Macropredatory ichthyosaur from the Middle Triassic and the origin of modern trophic networks. PNAS (2013).

Sander, P. M. et al. Early giant reveals faster evolution of large body size in ichthyosaurs than in cetaceans. Science (2021).

Motani, R. et al. Terrestrial Origin of Viviparity in Mesozoic Marine Reptiles Indicated by Early Triassic Embryonic Fossils. PLoS ONE (2014).

Motani, R. et al. A basal ichthyosauriform with a short snout from the Lower Triassic of China. Nature (2015).

Klein, N. et al. A new cymbospondylid ichthyosaur (Ichthyosauria) from the Middle Triassic (Anisian) of the Augusta Mountains, Nevada, USA. Journal of Systematic Palaeontology (2020).

Marchetti, L. et al. Leaving only trace fossils - the unknown visitors of Winterswijk. Staringia (2019).

Wintrich, T. et al. A Triassic plesiosaurian skeleton and bone histology inform on evolution of a unique body plan. Science Advances (2017).

Krahl, A. & Witzel, U. Foreflipper and hindflipper muscle reconstructions of *Cryptoclidus eurymerus* in comparison to functional analogues: introduction of a myological mechanism for flipper twisting. PeerJ (2021).

Chapter 2: New Life on Land

Fernandez, V. et al. Synchrotron Reveals Early Triassic Odd Couple: Injured Amphibian and Aestivating Therapsid Share Burrow. PLoS ONE (2013).

Qvarnström, M. et al. Tyrannosaurid-like osteophagy by a Triassic archosaur. Scientific Reports (2019).

Qvarnström, M. et al. Beetle-bearing coprolites possibly reveal the diet of a Late Triassic dinosauriform. Royal Society Open Science (2019).

Qvarnström, M. et al. Exceptionally preserved beetles in a Triassic coprolite of putative dinosauriform origin. Current Biology (2021).

Nesbitt, S. J. et al. The oldest dinosaur? A Middle Triassic dinosauriform from Tanzania. Biology Letters (2013).

Pol, D. et al. Triassic sauropodomorph dinosaurs from South America: The origin and diversification of dinosaur dominated herbivorous faunas. Journal of South American Earth Sciences (2021).

Griffin, C. T. et al. Africa's oldest dinosaurs reveal early suppression of dinosaur distribution. Nature (2022).

Zahner, M. & Brinkmann, W. A Triassic averostran-line theropod from Switzerland and the early evolution of dinosaurs. Nature Ecology & Evolution (2019).

Hofmann, R. & Sander, P. M. The first juvenile specimens of *Plateosaurus engelhardti* from Frick, Switzerland: isolated neural arches and their implications for developmental plasticity in a basal sauropodomorph. PeerJ (2014).

Fernández, O. R. R. & Werneburg, I. A new massopodan sauropodomorph from Trossingen Formation (Germany) hidden as '*Plateosaurus*' for 100 years in the historical Tübingen collection. Vertebrate Zoology (2022).

Olsen, P. et al. Arctic ice and the ecological rise of the dinosaurs. Science Advances (2022).

Chapter 4: Germany's Dinosaurs

Sander P. M. et al. Bone histology indicates insular dwarfism in a new Late Jurassic sauropod dinosaur. Nature (2006).

Lallensack, J. et al. Dinosaur tracks from the Langenberg Quarry (Late Jurassic, Germany) reconstructed with historical photogrammetry: Evidence for large theropods soon after insular dwarfism. Palaeontologia Electronica (2015).

Foth, C. & Rauhut, O. W. M. Re-evaluation of the Haarlem *Archaeopteryx* and the radiation of maniraptoran theropod dinosaurs. BMC Evolutionary Biology (2017).

Rauhut, O. W. M. et al. A non-archaeopterygid avialan theropod from the Late Jurassic of southern Germany. eLife (2019).

Taylor, M. P. A re-evaluation of *Brachiosaurus altithorax* Riggs, 1903 (Dinosauria, Sauropoda) and Its Generic Separation from *Giraffatitan brancai* (Janensch 1914). Journal of Vertebrate Paleontology (2009).

D'Emic, M. D. et al. Anatomy, systematics, paleoenvironment, growth, and age of the sauropod dinosaur *Sonorasaurus thompsoni* from the Cretaceous of Arizona, USA. Journal of Paleontology (2016).

Mannion, P. D. et al. The earliest known titanosauriform sauropod dinosaur and the evolution of Brachiosauridae. PeerJ (2017).

Chapter 5: Argentina—Where the Giants Live

Carballido, J. L. et al. A new giant titanosaur sheds light on body mass evolution among sauropod dinosaurs. Proceedings of the Royal Society B: Biological Sciences (2017).

Rauhut, O. W. M. & Pol, D. Probable basal allosauroid from the early Middle Jurassic Cañadón Asfalto Formation of Argentina highlights phylogenetic uncertainty in tetanuran theropod dinosaurs. Scientific Reports (2019).

Chapter 6: Myanmar—Trapped in "Liquid Gold"

Xing, L. et al. A mid-Cretaceous enantiornithine (Aves) hatchling preserved in Burmese amber with unusual plumage. Gondwana Research (2017).

Peñalver, E. et al. Ticks parasitised feathered dinosaurs as revealed by Cretaceous amber assemblages. Nature Communications (2017).

Xing, L. et al. Hummingbird-sized dinosaur from the Cretaceous period of Myanmar. Nature (2020). ARTICLE WITHDRAWN

Bolet, A. et al. Unusual morphology in the mid-Cretaceous lizard *Oculudentavis*. Current Biology (2021).

Dunne, E. M. et al. Ethics, law, and politics in paleontological research: The case of Myanmar amber. Communications Biology (2022).

Chapter 7: The Spinosaurs

Smith, J. B. et al. A Giant Sauropod Dinosaur from an Upper Cretaceous Mangrove Deposit in Egypt. Science (2001).

Ibrahim, N. et al. Semiaquatic adaptations in a giant predatory dinosaur. Science (2014).

Henderson, D. M. A buoyancy, balance and stability challenge to the hypothesis of a semi-aquatic *Spinosaurus* Stromer, 1915 (Dinosauria: Theropoda). PeerJ (2018).

Ibrahim, N. et al. Tail-propelled aquatic locomotion in a theropod dinosaur. Nature (2020).

Hone, D. & Holtz, T. Evaluating the ecology of *Spinosaurus*: shoreline generalist or aquatic pursuit specialist? Palaeontologia Electronica (2021).

Sereno, P. C. et al. *Spinosaurus* is not an aquatic dinosaur. eLife (2022).

Barker, C. T. et al. New spinosaurids from the Wessex Formation (Early Cretaceous, UK) and the European origins of Spinosauridae. Scientific Reports (2021).

Barker, C. T. et al. A European giant: a large spinosaurid (Dinosauria: The-

ropoda) from the Vectis Formation (Wealden Group, Early Cretaceous), UK. PeerJ (2022).

Chapter 8: More than Just Bones

Wang, X. et al. *Archaeorhynchus* preserving significant soft tissue including
probable fossilized lungs. PNAS (2018).

Bell, P. R. et al. The exquisitely preserved integument of *Psittacosaurus*
and the scaly skin of ceratopsian dinosaurs. Communications Biology (2022).

Vinther, J., Nicholls, R. & Kelly, D. A. A cloacal opening in a non-avian
dinosaur. Current Biology (2021).

O'Connor, J. K. et al. Ovarian follicles shed new light on dinosaur reproduction during the transition toward birds. National Science Review (2014).

Bailleul, A. M. et al. Confirmation of ovarian follicles in an enantiornithine (Aves) from the Jehol biota using soft tissue analyses. Communications Biology (2020).

Brown, C. M. et al. An Exceptionally Preserved Three-Dimensional Armored Dinosaur Reveals Insights into Coloration and Cretaceous
Predator-Prey Dynamics. Current Biology (2017).

Chapter 9: Wyoming—The Hell Creek

Schmitt, A. D. Auf der Dinosaurier-Baustelle: Zwischen Knochen und
Kettensägen. National Geographic (August 2019). Written in German.

Schmitt, A. D. Wie verschifft man ein Stück Dinosaurier-Friedhof? National Geographic (August 2019). Written in German.

Chapter 10: *Tyrannosaurus*—The Measure of All Things!

Marshall, C. R. et al. Absolute abundance and preservation rate of *Tyrannosaurus rex*. Science (2021).

Carpenter, K. & Smith, M. Forelimb Osteology and Biomechanics of *Ty-

rannosaurus rex. Mesozoic vertebrate life. Bloomington: Indiana University Press (2001).

Carbone, C. et al. Intra-guild competition and its implications for one of the biggest terrestrial predators, *Tyrannosaurus rex*. Proceedings of the Royal Society B: Biological Sciences (2011).

Holtz, T. R. Theropod guild structure and the tyrannosaurid niche assimilation hypothesis: implications for predatory dinosaur macroecology and ontogeny in later Late Cretaceous Asiamerica. Canadian Journal of Earth Sciences (2021).

Schroeder, K., et al. The influence of juvenile dinosaurs on community structure and diversity. Science (2021).

Carr, T. D. A high-resolution growth series of *Tyrannosaurus rex* obtained from multiple lines of evidence. PeerJ (2020).

Bell, P. R. et al. Tyrannosauroid integument reveals conflicting patterns of gigantism and feather evolution. Biology Letters (2017).

Padian, K. Why tyrannosaur forelimbs were so short: An integrative hypothesis. Acta Palaeontologica Polonica (2022).

Bhullar, B.-A. S. et al. Birds have paedomorphic dinosaur skulls. Nature (2012).

Ullmann, P. V. et al. Taphonomic and Diagenetic Pathways to Protein Preservation, Part I: The Case of *Tyrannosaurus rex* Specimen. MOR 1125. Biology (2021).

Paul, G. S. et al. The Tyrant Lizard King, Queen and Emperor: Multiple Lines of Morphological and Stratigraphic Evidence Support Subtle Evolution and Probable Speciation Within the North American Genus *Tyrannosaurus*. Evolutionary Biology (2022).

Carr, T. D. et al. Insufficient Evidence for Multiple Species of *Tyrannosaurus* in the Latest Cretaceous of North America: A Comment on "The Tyrant Lizard King, Queen and Emperor: Multiple Lines of Morphological and Stratigraphic Evidence Support Subtle Evolution and Probable Speciation Within the North American Genus *Tyrannosaurus*." Evolutionary Biology (2022).

Chapter 11: Movement Captured in the Stone: What Footprints Tell Us

Schanz, T. et al. Quantitative Interpretation of Tracks for Determination of Body Mass. PLoS ONE (2013).

Lockley, M. et al. Titanosaurid trackways from the Upper Cretaceous of Bolivia: Evidence for large manus, wide-gauge locomotion and gregarious behaviour. Cretaceous Research (2002).

Chapter 12: Birds—The Last Dinosaurs

Prum, R. O. et al. A comprehensive phylogeny of birds (Aves) using targeted next-generation DNA sequencing. Nature (2015).

Cloutier, A. et al. Whole-Genome Analyses Resolve the Phylogeny of Flightless Birds (Palaeognathae) in the Presence of an Empirical Anomaly Zone. Systematic Biology (2019).

Wiemann, J., et al. Dinosaur egg colour had a single evolutionary origin. Nature (2018).

Field, D. J. et al. Late Cretaceous neornithine from Europe illuminates the origins of crown birds. Nature (2020).

Koschowitz, M.-C. et al. Beyond the rainbow. Science (2014).

Foth, C. et al. New specimen of *Archaeopteryx* provides insights into the evolution of pennaceous feathers. Nature (2014).

Smithwick, F. M. et al. Countershading and Stripes in the Theropod Dinosaur *Sinosauropteryx* Reveal Heterogeneous Habitats in the Early Cretaceous Jehol Biota. Current Biology (2017).

Hu, D. et al. A bony-crested Jurassic dinosaur with evidence of iridescent plumage highlights complexity in early paravian evolution. Nature Communications (2018).

Chapter 13: The End of the Dinosaurs

During, M. A. D. et al. The Mesozoic terminated in boreal spring. Nature (2022).

Alvarez, L. W. et al. Extraterrestrial Cause for the Cretaceous–Tertiary Extinction. Science (1980).

Santa Catharina, A. et al. Timing and causes of forest fire at the K–Pg boundary. Scientific Reports (2022).

Nicholson, U. et al. The Nadir Crater offshore West Africa: A candidate Cretaceous–Paleogene impact structure. Science Advances (2022).

Field, D. J. et al. Early Evolution of Modern Birds Structured by Global Forest Collapse at the End-Cretaceous Mass Extinction. Current Biology (2018).

ACKNOWLEDGMENTS

When I was in seventh or eighth grade, my English skills were below average, and my English teacher told me that I would never learn how to speak proper English, let alone write a single concise sentence. Yet here I am. It is virtually impossible for other people to know what we are capable of until we do it. We all have skills and talents in us that are not apparent to others, and sometimes it just needs a little push and some encouragement from friends, family, and mentors for us to show the world what we are capable of. They may not know if we succeed, but they have faith in us and trust in our potential. That is why I would like to thank those who made it possible for me to write this book. This is most certainly not a complete list—there may be a few people whom I forgot to mention, and I apologize for that in advance.

I am particularly grateful for all the help and support I received from Martin Sander in Bonn, who was my supervisor for my diploma thesis. He recognized my potential and gave me a chance to prove myself, but I also learned so much from him that I am now myself able to pass on my knowledge to others. Equally important for my career in paleontology is Roger Benson, whom I worked for as a research assistant in Oxford. I can't overstate how lucky I feel to have had the opportunity to work with one of the most brilliant minds I have ever encountered in my life. If it wasn't for this job, many other consecutive career steps would not have been possible. I owe gratitude to Daniel Field for offering me a doctoral research position in Cambridge. Had it not been for this position, I would have never been interviewed by Frank Thadeusz from the news journal *Der Spiegel*. Thanks to him, people started taking an interest in my dinosaur content on various social media platforms, and so it was him who pointed the publisher DTV in my direction. Here, I would like to thank Laura Weber, who offered me the opportunity to write a book about my favorite topic: dinosaurs. And Andrea Seibert, who did a tremendous job at presenting my popular science book to a list of international publishers at the London Book Fair. I am very thankful to Hanover Square Press for giving me the opportunity to present my book to an English-speaking audience and thus sharing my love for dinosaurs with people across the globe. At Hanover Square Press, Eden Railsback and Peter Joseph did a phenomenal job with helping me convert this book into a presentable English version. They both have been supportive and incredibly encouraging from the moment we started working together.

And while the book is out now and I am proud of my accomplishment, the book came about in a time when I was struggling with severe personal problems, and had it not been for friends like Sabrina Müller and the wonderful Maryness Honoratus, I would not have been able to push through and overcome all these challenges. I would also like to thank Ben Marshall from Hughes Hall College for providing mental support, and Philip Isaac from King's College, who helped get me back on my feet. In the end, it is always the success that other people see, and never the struggles we encounter along the way.

A special thanks goes out to Steve Brusatte, who has always encouraged me to continue to pursue my career in paleontology and to write a book, and to Thomas Holtz, who has been very supportive during the writing process of this book by providing valuable information and insight into the anatomy and physiology of *Tyrannosaurus*. I would also like to thank my former housemate Davinder, who helped me with an earlier translation of two chapters of this book, albeit his translation ultimately did not end up in the final product. Nevertheless, I acknowledge and appreciate his effort. Of course, I would like to thank Lars Schmitz and Georg Oleschinski for letting me use their photos for this book!

Lastly, I would like to thank my family for making me the person I am today. I owe everything to my mother and my father, who may not have always understood me but who always made sure I felt loved unconditionally and always accepted me for who I am, despite being difficult or awkward at times. I would also like to thank my brother, Michael, and my sister, Sonja, for keeping up with me, despite being the most annoying youngest sibling imaginable.

Most importantly, all my love goes out to my beautiful children, Maximilian and Elisabeth, who are the most precious and wonderful human beings imaginable. They are in my heart always and they are the apples of my eye and the reason and motivation behind my achievements. I hope they feel loved unconditionally by their dad, as I feel loved by my parents.

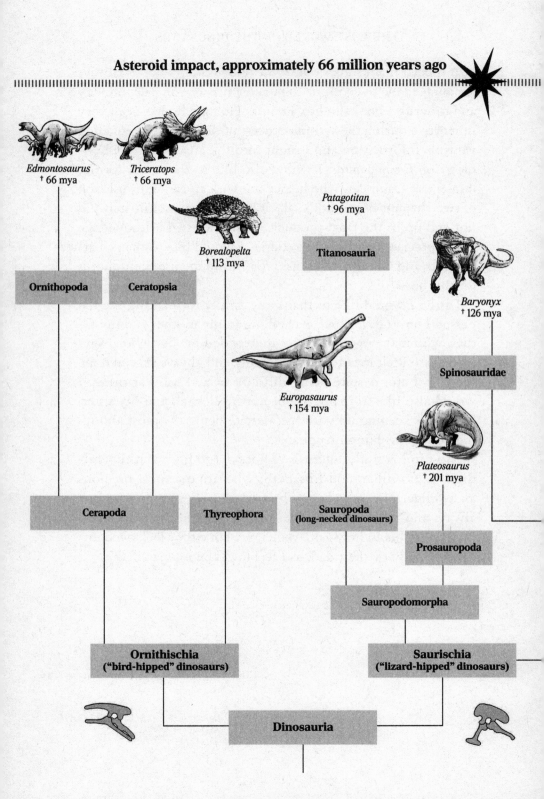

Asteroid impact, approximately 66 million years ago

Edmontosaurus
† 66 mya

Triceratops
† 66 mya

Borealopelta
† 113 mya

Patagotitan
† 96 mya

Titanosauria

Baryonyx
† 126 mya

Ornithopoda

Ceratopsia

Europasaurus
† 154 mya

Spinosauridae

Plateosaurus
† 201 mya

Cerapoda

Thyreophora

Sauropoda
(long-necked dinosaurs)

Prosauropoda

Sauropodomorpha

Ornithischia
("bird-hipped" dinosaurs)

Saurischia
("lizard-hipped" dinosaurs)

Dinosauria

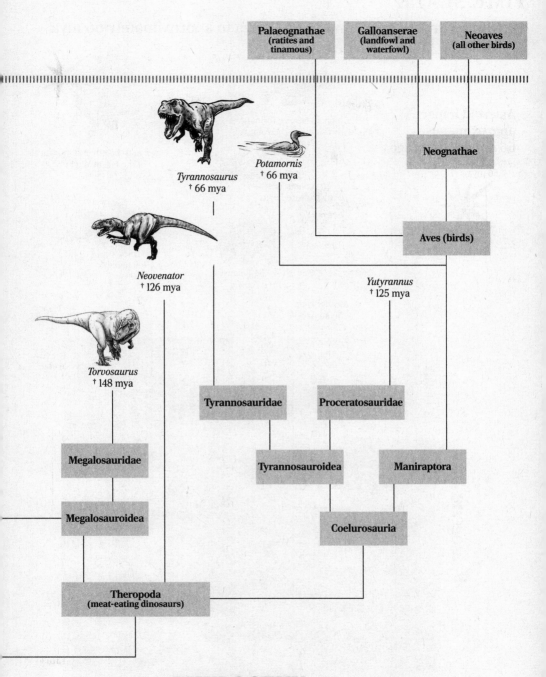

PHYLOGENY

Dinosaur Phylogeny until the asteroid impact, approximately 66 million years ago

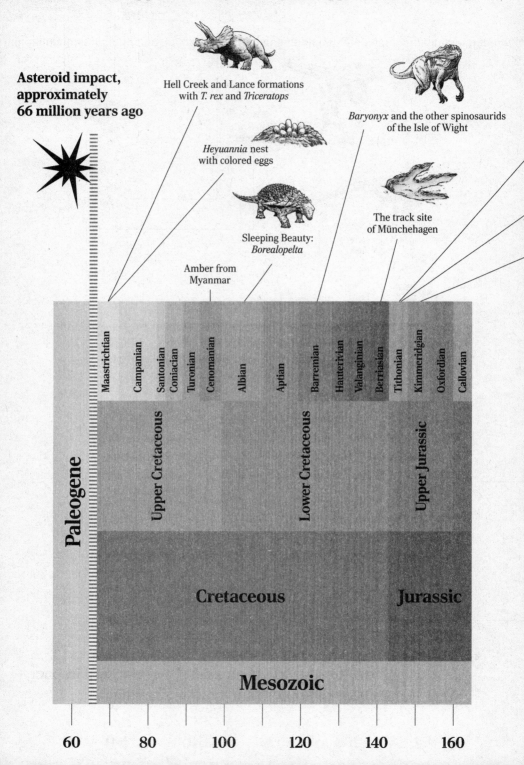

TIME SCALE
Time Scale from approximately 252 mya to approximately 66 mya

Asteroid impact, approximately 66 million years ago

Hell Creek and Lance formations with *T. rex* and *Triceratops*

Baryonyx and the other spinosaurids of the Isle of Wight

Heyuannia nest with colored eggs

The track site of Münchehagen

Sleeping Beauty: *Borealopelta*

Amber from Myanmar

Maastrichtian

Campanian

Santonian
Coniacian

Turonian

Cenomanian

Albian

Aptian

Barremian

Hauterivian
Valanginian

Berriasian

Tithonian

Kimmeridgian

Oxfordian

Callovian

Paleogene

Upper Cretaceous

Lower Cretaceous

Upper Jurassic

Cretaceous

Jurassic

Mesozoic

60   80   100   120   140   160

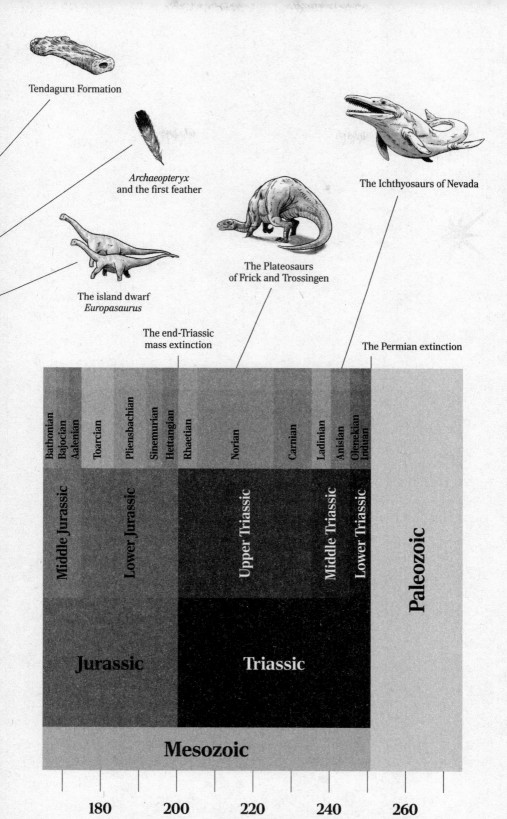

Tendaguru Formation

Archaeopteryx
and the first feather

The Ichthyosaurs of Nevada

The Plateosaurs
of Frick and Trossingen

The island dwarf
Europasaurus

The end-Triassic
mass extinction

The Permian extinction

Bathonian
Bajocian
Aalenian
Toarcian
Pliensbachian
Sinemurian
Hettangian
Rhaetian
Norian
Carnian
Ladinian
Anisian
Olenekian
Induan

Middle Jurassic

Lower Jurassic

Upper Triassic

Middle Triassic

Lower Triassic

Paleozoic

Jurassic

Triassic

Mesozoic

180   200   220   240   260

INDEX

first remains of, found, 76–77

global discoveries of, 82

juveniles, 80–81

nickname of, 77

as prey, 79

sites in Europe with, 76, 77–78, 79, 80–81

playa strata, 79

plesiosaurs, 53–55, 54(illus)

plumage. *See* feathers

poachers, 49

Polacanthus, 161

Potamornis, 176(illus)

primary producers, 41

Prince Creek Formation (Alaska), 214

Procolophonichnium, 53

prosauropods, 83–84

protofeathers, 248

Psittacosaurus, 166–68, 167(illus)

Q

Qvarnström, Martin, 70, 71–72

R

Rauhut, Oliver, 109, 124–26, 128–30

rebbachisaurids, 96

Red Nose Point (Augusta Mountains), 28

regurgialites, 70

reproduction

birds, 134, 241–42

dinosaurs, 203, 208, 242–43, 248

egg coloration, 241–42, 243–44

enantiornithines, 168, 169

feathers and, 249

Jeholornis, 168–69

limit to egg size, 237–38

medullary bone tissue and, 220–21

paleognath birds, 236–37

Triceratops, 180

research, funding and publishing, 136–37

retrodeformation of bones, 141

Rhaeticosaurus, 54

rhinoceros, dwarf, 103

rhynchosaurs, 52

Riggs, Elmer, 113

Riojasaurus, 82

Riparovenator, 163

Royal Tyrrell Museum (Drumheller, Canada), 170

S

Sacrison, Stan, 195

Sander, Martin, 38(illus)

affinity of *Omphalosaurus* to ichthyosaurs, 34

basic facts about, 32

Europasaurus and, 102

field designations of skeleton and, 46–47

juvenile *Plateosaurus* and, 80–81

prediction about proto-birds from Langenberg quarry, 109